John Postgate, FRS, is Emeritus Professor of Microbiology at the University of Sussex, where he was also Director of the Unit of Nitrogen Fixation. He was educated at Kingsbury County School, among others, and Balliol College, Oxford, where he took a first degree in chemistry before turning to chemical microbiology. He then spent fifteen years in government research establishments – studying mainly the sulphur bacteria and bacterial death – before moving to the Unit at Sussex, where he spent the next twenty-two years. He has held visiting professorships at the University of Illinois and Oregon State University and has been President of the Institute of Biology and of the Society for General Microbiology.

He is the third Professor John Postgate: the first his great-grandfather taught medicine at Birmingham University, the second his grandfather taught classics at Liverpool University. His other grandfather was George Lansbury, the Socialist leader, and his father was Raymond Postgate, the historian and gourmet. Long ago John Postgate led the Oxford University Dixieland Bandits (on cornet), and he is known as a jazz writer. He and his wife, who read English at St Hilda's College, Oxford, have three grown-up daughters.

Since the dawn of life on Earth, the world has been gradually transformed by living things into a comfortable home for plants, animals and ourselves. But many harsh and seemingly inhospitable places remain, and it is the inhabitants of such places, mainly invisible microbes, that reveal the remarkable potential and resilience of life itself. How do microbes survive, even flourish, in superheated water or supercooled brine; at enormous pressures; without air; amid poisons? And what part do, and did, they play in making the Earth hospitable?

In this fascinating account, for lay readers, John Postgate, one of Britain's leading microbiologists, tells of the diverse adjustments microbes have made to apparently impossible habitats. Modern understanding provides new clues to the origin and evolution of terrestrial life, offers glimpses of how life might have established itself elsewhere in the universe, and raises profound questions about death, sensation and individuality – as well as illustrating the often muddled pathways of scientific progress.

Related title

Microbes and Man, 3rd edition
JOHN POSTGATE FRS

JOHN POSTGATE FRS

The Outer
Reaches
of Life

CAMBRIDGE
UNIVERSITY PRESS

Published by the Press Syndicate of the University of Cambridge
The Pitt Building, Trumpington Street, Cambridge CB2 1RP
40 West 20th Street, New York, NY 10011-4211, USA
10 Stamford Road, Oakleigh, Melbourne 3166, Australia

First published 1994
Canto edition 1995

Printed in Great Britain at
the University Press, Cambridge

A catalogue record for this book is available from the British Library

Library of Congress cataloguing in publication data

Postgate, J. R. (John Raymond)
The outer reaches of life / John Postgate. – 1st ed.
p.   cm.
ISBN 0 521 44010 6
1. Extreme environments–Microbiology.   I. Title.
QR100.9.P67   1994
576–dc20   93-11579   CIP

ISBN 0 521 44010 6 hardback
ISBN 0 521 55873 5 paperback

RO

Contents

Preface

Almost a quarter of a century ago I wrote a book, *Microbes and Man*, in which I gave an account of the world of microbes – bacteria, viruses, protozoa and the like – and described their impact on our lives, our economy and our society. It was written primarily for non-scientist readers and has been a success, I am happy to say: translated into half a dozen languages and now into its third edition.

A dozen years after it came out my late aunt came across it. She was Dame Margaret Cole, classicist, writer, educationalist and Fabian socialist. Surprised that her nephew had written a paperback, she dutifully read it.

My aunt belonged to the generation of educated people which took it for granted that scientists were essentially uneducated, as well as barely literate. (I would not deny that a brief scan of professional scientific journals is likely to confirm her view, but how did she know?) Having read my book, she was moved to write to me – a thing she rarely did – expressing, with disarming frankness, her pleased surprise that she understood and even enjoyed considerable portions of it. She added that it would have been even better if I had left out the chemistry: those formulae and equations, with their lines, arrows and rows of capital letters – they put her off.

I felt humbly complimented. But I have worried ever since, on and off, about her response to equations and formulae. It is widely shared. Yet equations (mathematical or chemical) and formulae, like diagrams and graphs, are more than scientific shorthand; they can be beautiful when you understand the picture they reveal. Nevertheless, they can also be a deterrent for those who are not tuned in to such things. How, then, without formulae and diagrams, does one explain the glorious inwardness of biology, when so much of it is rooted in chemistry, particularly when the biological entities are microbes?

Should one shrug one's shoulders? Shed a tear, perhaps, for that huge Other Culture in our society, comprised of all those who were never told, or have forgotten, the basics of chemistry, physics and biology? Of course not; there is not time. For it is they who are daily obliged to make decisions on problems where scientific understanding is essential, decisions on matters which range from the merits of nuclear energy to the choice of baby food.

'Scientific understanding', I said, not 'scientific knowledge'. The distinction is important. Science has grown so big that no brain can encompass even a moderate fraction of the vast mass of knowledge which is preserved in print and on disk. Of course, good scientists have to have good memories, but the era of the polymath is over. Most scientists specialise in subdivisions of their main subject, and have little real knowledge of other areas, let alone of other disciplines. But scientific understanding is another matter. Reading or hearing about science, one allows oneself to forget, in self-defence, most of the detail on offer, but experiences new insights, recognises new relationships among things, and learns new patterns of thought and logic. Our perception of the world we live in is for ever changed, and our awareness of our place in it is enhanced.

Scientific understanding changes lives. It is – or can be – quite as valuable, formative and enjoyable a cultural experience as absorbing a great work of art or literature.

The day will come when the elements of science, and its logic, are as much a part of human culture as reading, writing and arithmetic. But that day is still a generation or two away, and meanwhile science races on, spawning new technologies and raising new ethical problems the while. And as it advances, it also illuminates ever wider vistas of our cosmos, from its tiniest components to the whole universe. Must the cultural delight in the understanding that science brings be closed to over half the populace?

I do not think so, and in this book I have tried to convey, to non-scientist readers, something of the way in which our understanding of the largely invisible world of microbes is giving us new slants on life itself. Avoiding formulae, maths, technical terms and graphs as much as I could, I have tried to display the glimpses microbes offer of exotic ways of life, some of which were probably dominant at various stages of this planet's infancy, and some of which may well have become the norm among creatures elsewhere in the universe.

John Postgate
Lewes, 1993

Some readers will find it helpful to read Chapter 1 first, because it includes a brief introduction to microbes in general. Thereafter there is no need to take chapters consecutively – even though there is a logic in their order.

Acknowledgements

The hobyahs, whose visual comments on the chapters so much enhance my book, are imp-like creatures, in no way resembling microbes, visualised by John D. Batten for one of the classic nineteenth-century collections of traditional fairy stories assembled by Joseph Jacobs. I thank Sue Shields for the vignettes and congratulate her on the way in which her adaptations retain the manic vivacity of Batten's original beasties.

I am also grateful to the Editors of the following publications for permission to include portions of previously published writings: *The Student Book 1988/89 et seq.* (Macmillan) for part of Chapter 1; *New Scientist* for parts of Chapters 3, 7, 10, 16 and 17; *Times Higher Education Supplement* for parts of Chapters 11, 12 and 14.

Finally, I thank my wife, Mary Postgate, for reading, criticising and improving every chapter.

1
Microbes and terrestrial life

Microbes, collectively, are the most versatile of living things. They possess a range of abilities that is enormously wider than we encounter among higher organisms; in a sense they define the biochemical limits of terrestrial life.

That assertion needs explaining, but first I must provide a quick run-down on what microbes are, and do.

Microbes are tiny living things, almost always single cells, and they can only be seen satisfactorily through a microscope. A few, such as amoebae and certain sulphur bacteria, are just discernable, but nothing more, with the naked eye, and most microbes are visible under a decent microscope; but viruses are so small that even the most powerful optical microscope is inadequate and scientists have to use an electron microscope to see them. Some microbes are like very simple plants – yeasts, moulds and certain algae are examples. Others, such as the amoeba, are tiny primitive animals. But the majority, the bacteria, are neither plants nor animals; they constitute a different 'Domain' of living things. (Actually, that Domain is divided in two, the Domains called Archaebacteria and Eubacteria, and no-one is quite sure where viruses fit in. But these are technicalities.)

Microbes are everywhere: in the air, in soil, in water, on our skin and hair, in our mouths and intestines, on and in the food we eat. They were first seen by a Dutch microscopist, Antoni van Leeuwenhoek, in the mid seventeenth century; in letters to London's Royal Society he described the 'animalcules' he had seen, with his primitive microscope, in samples of water, soil and body fluids. He included drawings which have since been identified as representing known species of microbe. Mostly he saw the larger types of microbe – protozoa and algae – as well as tiny worms and such higher organisms, but on 26 December 1676, looking at water in which pepper had been steeped overnight, he saw, for the first time, manifestly living creatures which were several times tinier than those he had described earlier. We now know them as bacteria. Boxing Day, 1676, was the birthday of the science of Bacteriology! It is unusual to know the date on which a new science originated with such precision, but Leeuwenhoek was meticulous in recording such details in his letters to the Royal Society.

Microbes influence our lives in a tremendous variety of ways. They make the soil fertile; they clean up the environment; they change, often improve, our food; they make vitamins for us inside ourselves; they can protect us from other, undesirable, microbes. Yet most people are scarcely aware that they exist and, as 'germs', they have a bad press. There is a simple reason for this: a few kinds of microbes cause disease, a few spoil food, or destroy valuable materials. And the occasions when such misfortunes happen are about the only times most people notice microbes at all.

Some bacteria do indeed cause illnesses, such as whooping cough, scarlet fever, typhoid and so on; illnesses that have become rare today because of the effectiveness of immunisation, drugs and antibiotics. Though a few awkward bacterial diseases persist, and poverty, war and AIDS have revived some that we thought were conquered, it is still fair to say that the viruses are today's major medical problem. They are infinitesimal fragments of barely living material which can cause diseases ranging in severity from the common cold to AIDS. But the war against even those creatures is slowly being won. Smallpox has been eradicated using a combination of hygiene and vaccination; polio is in retreat; modern genetic engineering is providing new ways of attacking other virus diseases. Microbes are also important in animal and plant diseases, but today some of those that attack plants can actually be recruited to add new and useful genes to plants, creating new varieties. And as well as causing disease among higher organisms, certain microbes can attack non-living material: they can corrode concrete and iron pipes, and spoil leather, wood, paper and even glass or plastics; the troubles they can cause in oil technology, mining, machining and even film-processing would fill a book, not to mention the part they play in food spoilage and water pollution.

But just as important are the beneficial, useful aspects of the microbial world. The crucially important part they play in sewage treatment and waste disposal may seem mundane and boring, but it is fundamental to our social well-being; microbes also clean up our fouled lakes, rivers and beaches, even deal with exotic industrial effluents; they decompose plant material, corpses and excreta to renew soil fertility. They add essential nitrogen to soil, help plants to get at phosphate, and keep up the supply of sulphur compounds which plants and animals need. In all, by recycling the detritus of plants and animals they constantly renew the supplies of oxygen, carbon dioxide, nitrates and even water on which all life on this planet depends. So vital are microbes to the lives of higher organisms that our own nutrition depends on them, and some animals – sheep, cattle, termites, for example – carry microcosms of specialised microbes around in their guts, to digest materials which the animals alone could not tackle. Plants, too, often have 'helper' microbes in their roots or leaves; symbiosis is the Name of the Game in biology.

All this is textbook stuff. The impact of microbes on our health and well-being, on our society, on our economy and industry, is an absorbing topic and has been the subject of several books, technical and 'popular'. But as far as this chapter – indeed, this book – is concerned, there is a deeper fascination about microbes which stems from the way in which they do these things. Which brings me back to my opening remark about 'all the processes of which living cells are capable'.

Consider dogs and fish. Though they live in very different environments, their life processes are actually very similar: they breathe oxygen, they eat organic food, they reproduce sexually, they are made up of enormous numbers of rather similar cells, they live out their lives and then they die. All this is true of all animals, from flatworms to people. And even plants are really much the same fundamentally, except that they mostly do not eat organic food; instead they use solar energy to make their organic matter from carbon dioxide. What makes microbes, especially bacteria, so different is the diversity of their life processes at this very basic level. Thus, many types of bacteria need no air; either they decompose, rather than burn, organic food, or they get oxygen from oxygen compounds such as sulphates or nitrates, and not from the air. Types of bacteria exist which keep their vital processes going by transforming iron compounds or sulphur. Some kinds thrive in boiling water; others in sub-zero brines. Some sense magnetic fields; some grow at huge hydrostatic pressures. And it seems that most bacteria are potentially immortal: they die only if some stress kills them (you can get into quite a philosophical tangle here if you have a taste for that sort of thing). Spores of certain bacteria survive dormant for thousands of years; other bacteria are so fragile that they die seemingly as soon as you look at them.

There have been living things on this planet for about 3½ billion years, and for most of that time they were creatures rather like today's microbes. In the beginning they lived in a pretty harsh world: no oxygen, lots of radiation, big fluctuations in temperature, humidity and salinity, abrupt environmental changes due to earthquakes, vulcanism, drought and inundation. Glaciation, too, became part of their lives later on. The microbial world learned, perforce, to survive, even to flourish, in tough conditions, making use of every geochemical resource that could be turned to advantage, exploiting every favourable change as rapidly as possible. And in doing so, the microbes began to change the world; they started a process of biological transformation of the planetary environment which continues to this day.

Something like 1½ billion years ago, when the chemical turmoil of earlier aeons had quietened down considerably, certain microbes learned to perform the sort of photosynthesis that plants conduct today. By this I mean that they not only learned to use solar energy instead of energy derived from organic food, but they coupled the process to releasing gaseous oxygen from water. That oxygen altered the composition of the atmosphere permanently, and paved the way for the emergence of oxygen-breathing creatures. And about a billion years ago air-breathing organisms arose which were composed of many cells, cells which had diverse, though co-operative, functions. These creatures, as they evolved into the animals and the plants, gradually took over the planet. One kind of life, one sort of biochemistry, became dominant among living things (I amplify the story of life's early history in Chapters 14 and 17).

Evolution progressed, and some of these higher organisms developed remarkable abilities. Co-operation among the component cells of animals, especially, became progressively more sophisticated and in time allowed the development of complex skills ranging from web-spinning and nest-building to the use of tools and machines, and the development of language and conceptual thought. But the life processes that fuelled the cells of these complex organisms changed very little; they stuck to the oxygen-based biochemistry which is common to both plants and animals.

Only the microbes, especially the bacteria, retained most, if not all, of the diverse potentialities of terrestrial life. It is this diversity that underlies the fascination of which I wrote earlier, with its intimations of how things might have been – or may well be elsewhere in the universe.

2

Some like it hot

Getting cooked

We all know what happens when you boil an egg. The gooey mass inside the shell starts to coagulate and, within minutes, it sets to a gastronomically pleasing consistency. Much the same happens when you poach a piece of fish, except that it takes a little longer. One thing is certain: if either the piece of fish or the egg had been alive, it would have been dead within seconds. (In today's infertile eggs, by the way, there is actually nothing which can be called alive.) Living material of the kind we normally encounter remains alive for almost no time at the boiling point of water: it gets cooked. So to me, one of the strangest features of life, viewed as a whole, is that life forms exist which are able to flourish at high temperatures, some thriving at, and even above, the normal boiling point of water.

Let me put the matter in perspective. We humans keep our bodies at 37° Celsius. (When I was young, people used to say 98.4° Fahrenheit, so let me remind older readers that 37 °C and 98.4 °F are the same, and add that Celsius is as nearly the same as the older Centigrade as makes no odds: it runs from 0°, the freezing point of water, to 100°, its boiling point. Degrees Fahrenheit, in which the same range was from 32° to 212°, were always awkward to remember, as well as difficult to do arithmetic with; as far as temperatures are concerned, I rejoice in decimalisation, so temperatures in this chapter will henceforth be in degrees Celsius.)

Back to the perspective. Our body temperatures can fluctuate by a couple of degrees around 37°, and we feel ill if they do. If they move much outside that range for long, we die. Because our temperature is so important to our well-being, our biological machinery for regulating it is highly efficient. For example, one can sit in a hot bath, say at about 48°, for quite a long time, coming to no harm because one's blood circulation increases, to conduct the heat away to cooler parts of the body. Provided exposure is not too prolonged, one simply ends up looking a little flushed. Conversely, one can bathe in the cold sea at a British 15°, and survive – and enjoy if you are that crazy – brief immersion at 4°, the temperature of ice-water; again, the circulation accelerates, this time to keep one's outer regions warm. In addition, our metabolic rate responds rapidly to temperature, so that we burn, as appropriate, more or less of our food reserves.

Other warm-blooded creatures, mammals and birds, keep themselves warm in similar ways, their healthy temperatures ranging, according to species, over a couple of degrees either side of our 37°. Some mammals can hibernate, which means they can 'switch off' their thermal controls and may cool down nearly to freezing point without coming to harm. Cold-blooded animals, like all plants, remain more or less at the external temperature, showing ever-decreasing vitality as the temperature lowers. Dormant forms of life, such as seeds and spores, often survive periods at a few degrees below freezing.

I shall write more about cold in Chapter 3. Here I am concerned with heat, and the message is that the living things that we see around us, no matter where we live on this planet, live their active lives in a temperature range from 0° to about 48°. There are specialised animals, such as certain insects and crustaceans which live around hot springs, which seem able to survive, even to flourish, at a few degrees warmer – their limit seems to be about 50°, but give or take that couple of degrees, 48° represents the upper temperature limit of ordinary, visible plant or animal life.

I ought to mention here that a few kinds of bacteria can form remarkably heat-resistant bodies called spores. They can be very tough, able to survive freezing, drying and most disinfectants, as well as heating, sometimes even boiling. But they are dormant: virtually none of the ordinary life processes of living cells take place within them, and their quota of vital constituents, which is minimal, is inactive, protected in a jellied mass within their resistant coat.

An upper limit of around 48° for active life makes general biological sense. After all, there are few places on Earth where the temperature goes much above 45° for long, and the few tropical regions where this does happen are generally so arid as to exclude all but dormant forms of life. So there has been no compelling reason why most living things should have learned to live at high temperatures during their evolution. But there do exist places which are consistently both hot and wet: geothermal springs. These are locations, usually associated with volcanically active areas, where deep subterranean water, heated by the earth's molten core, finds a leak in the earth's crust and escapes. Examples include the hot springs of Yellowstone National Park in Montana, USA, and North Island, New Zealand; or the geothermal ponds of Iceland; even our modest British spas at Bath or Leamington. Such springs occur under the sea, too – I shall return to them later.

Hot springs have existed for millions of years. Why have plants and animals not evolved so as to colonise them?

There are lots of reasons. One is that ordinary living matter, at temperatures much above about 50°, is cooked – like the egg I wrote of. Protoplasm, the living fluid of all living cells, is a solution in water of large, delicate molecules called proteins, with a few minerals and small molecules. A hen's egg comprises one live cell – the ovum – and a huge mass of protein which would, if the egg were fertile, feed and cushion the developing embryo. When you boil it, the proteins coagulate. What this means chemically is that heat has destroyed the proteins' molecular structures: they are no longer the molecules that they were naturally. The technical term is that they have been 'denatured', and the undamaged proteins are referred to as 'native' proteins.

In most ways, native and denatured proteins are chemically unchanged. Proteins, of all sorts, are made up of chains of much smaller molecules called amino-acids. These are compounded of carbon, hydrogen, oxygen, nitrogen and occasionally sulphur atoms, and the atoms are arranged in a special way which enables amino-acids to stick together in chains. About twenty kinds of amino-acid are found making up proteins, and the gross differences which biochemists – and cooks – observe among proteins depend primarily on the variety of amino-acids constituting them, and their permutations and combinations along the amino-acid chain. But the fine differences, those that matter to living things, depend on the three-dimensional shape that an amino-acid chain assumes. For one thing, it is never straight; the preferred shape for a chain of amino-acids is twisted. So a typical molecule of a native protein will be a sort of bundle, with stretches in which the chain is twisted into a coil like a telephone lead (a 'helix' is the technical term), the coil itself following a curved course; then there might be a jagged stretch to turn a corner, leading to another coiled section. The whole is folded, sometimes into a spherical or oval shape. Though the majority of proteins are like this, some almost linear protein molecules are known – in hair and muscles, for example.

Native proteins have 3-D shapes which are central to their individuality and, with a few exceptions, their shapes are sustained by water. In general, molecules of water, which pack out odd spaces within and surrounding the molecular bundle, interact with the amino-acids composing the chains and sustain the protein's 3-D shape. The water molecules also keep the majority of proteins in a form enabling them to dissolve in water. Protoplasm contains three or four thousand subtly different kinds of protein molecules, capable, principally because of their special 3-D structures, of performing all the biochemical processes which add up to life: they are called enzymes.

What happens when you boil an egg, or cook any other protein for that matter, is that the heat agitates all the molecules indiscriminately: water, protein, minerals and everything else. This activity disrupts the 3-D shapes of the proteins: folded zones get shaken apart, the two ends of the chain become loose and the whole structure begins to open up. If a protein is not too badly unfolded, as after a short heat stress, it may remain capable of reassembling into its native shape when cooled, but this happens with only a few proteins. Prolonged heat scrambles the vast majority. The first thing to happen is that they lose their solubility in water: they coagulate, as in a boiled egg. The denatured proteins are made up of the same chains of amino-acids as they were when they were native – only rarely are these chains broken, except by prolonged or severe heating – but their subtle 3-D structure has become an irreversible, random tangle. And, of course, their biological activity has vanished: if they were part of an active, living thing, that thing is dead.

Moreover, you do not need to boil an egg to coagulate its proteins. Even at 65° it will set, though it takes much longer. And even at 50° the 3-D structure of some proteins is badly shaken, and becomes permanently disturbed quite soon.

That is one reason why the familiar denizens of this planet have an upper temperature limit of around 48°. Here is another. Cells are surrounded by membranes – they have to be, or their contents would leak. But these membranes are rather special in that they let things into the cell which it needs – nutrients and oxygen, for example – and also let waste materials out. The membranes are made partly of protein, but they are also waxy – fats are involved in their make-up. And, as well you know, when you heat fats, they melt. So when a cell becomes overheated, parts of its membrane tend to melt, leaks appear and, if heating goes on for too long, the cell dies. Microbes such as yeasts and bacteria can, if a heat stress is gradual enough, readjust the waxiness of their membranes a bit so as to cope with elevated temperatures, but only within narrow limits.

Denatured protein and melting membranes provide two reasons why 48° is a sort of temperature limit for the living things we see around us. There are other reasons. Some of the small molecules involved in generating energy break up rapidly in hot water, and the machinery whereby the information encoded in the cell's genes is processed and made use of proves very heat-sensitive. The upshot is that even mildly excessive heat can stress, even destroy, the machinery of living cells in a multitude of ways.

But living things are resilient, and it will come as no surprise to know that cells of all kinds have developed ways of coping with heat stresses, mild ones at least. I mentioned that some microbes can adjust the waxiness of their membranes so that they melt less easily. Most cells can counter heat stresses on their proteins. A cell, becoming overheated, will begin to make new proteins, the so-called 'heat-shock' proteins. These substances are not only rather heat-stable in themselves: they also tend to wrap themselves around other proteins and to stabilise them protecting them from denaturation. They have been given the rather charming name of 'molecular chaperones', or 'chaperonins'. They probably protect more complex cell structures, too, but – and it is an important but – their effect is usually short-lived and only works up to 5° or so above the organism's normal temperature limit.

Avoiding getting cooked

However, among the living things that we cannot see, except under the microscope, things are different. Since the early decades of this century, microbiologists have known of 'thermophiles': types of bacteria which flourish at 55 to 65°. They turn up in hot-water systems, or in the sugar industry (where molasses has to be kept hot during sugar refining). They also inhabit the interiors of silage towers and compost heaps, which get quite hot as the vegetable matter ferments. Thermophile means 'heat-lover'; they do not grow, or they grow only very slowly, at ordinary temperatures. Laboratory studies in the mid century established that such thermophiles have solved the protein and membrane problems: they possess specially heat-stable proteins and have high-melting fats in their membranes. Indeed, certain enzymes from such bacteria – those that disperse fats, or break down proteins, or solubilise starchy materials – have long been exploited in 'biological' detergents because, unlike comparable enzymes from ordinary organisms, they do not mind if the washing water is hot. Some 'hot' thermophiles would even grow at temperatures around 70°, but until about a couple of decades ago that seemed to be the upper temperature limit for all terrestrial life.

That, too, seemed to be a logical limit, for this reason. As most people know these days, the genetic blueprint for living things is carried in a molecule called DNA. (The initials DNA stand for deoxyribonucleic acid, but I can't keep writing that.) DNA is the stuff of which genes are made. All living cells, in all organisms (except, curiously, the red blood cells of mammals), carry a copy of that blueprint, a DNA molecule, and this, too, is a long chain of smaller molecules, The small molecules are quite different from amino-acids, and they are not all of one type, but their chains also form coils, and DNA is made up of two such coils wound around each other – as if two telephone leads had been neatly laid alongside each other, coil by coil. The two coils form the 'double helix' of scientific and TV fame. The two coils separate, and then but momentarily, only when the cell is reproducing itself. Yet the two strands are actually rather weakly stuck together and, when DNA is heated, they come apart. This so-called 'melting' of DNA's structure happens at between 65 and 75°. In contrast to proteins, the two strands come back together into the right relative positions when cooled. However, for a living cell to work, to use the information encoded in its genes, the DNA chains have to be double: melted DNA is useless.

Thirty years ago Dr. Ellis Klemperer of Bethesda, Maryland, USA, took up this point in an article for the American magazine *Science*. He concluded that 73° was the upper limit for life. Above that temperature, DNA would unravel itself, and then all life processes would become scrambled, would run-down and cease.

At the time it seemed that he was right. But within a few years the known upper limit of bacterial life began to creep upwards,

as new creatures able to grow at temperatures closer and closer
to boiling point were discovered. Around 1969, for example, Dr.
Thomas Brock of the University of Indiana, USA, investigating
microbes in the hot springs at Yellowstone National Park, USA,
found several new bacterial denizens which broke the 73° barrier.
One, which he named *Thermus aquaticus*, took the record (for the
time being) by growing at around 80°. But the big surprise came
because of a dramatic advance in a wholly different area of knowledge: geography.

Submarine exploration, using deep-sea submersibles, revealed
the existence of hot springs in the ocean bed. They are called
hydrothermal vents, and they lie along submarine rifts where
tectonic movement of the Earth's crust is slowly separating
two continental masses. There is a rift of this kind, called the
Galapagos Rift, some 2½ kilometres deep in the Pacific Ocean,
and this was examined in the late 1970s using a submersible
called *Alvin* (which belongs to the Woods Hole Oceanographic
Institute, Massachusetts, USA), which could take water samples.
The water which emerges from such springs has been heated
geothermally while within the Earth's crust but the hydrostatic
pressure down there is so high that it does not boil: it superheats,
remaining liquid up to 350°. Though it cools, often considerably,
on its way to the vents, temperatures ranging from the low 300s
and mid 200s down to 25° have been recorded at their orifices.
Of course, even the hottest emergent water cools rapidly as it
meets a chilly ocean at about 2.5°, but in the warm, turbulent
zones close to the hot vents may be found bacteria which flourish
at 100 to 110°. More ordinary, non-thermophilic bacteria are
found in the cooler zones, providing sustenance for quite a community of worms, clams and crabs (also non-thermophilic, of
course).

Considerable excitement was generated in the early 1980s by
a report, in the scientific journal *Nature*, that bacteria had been
detected growing right within the effluent of a hydrothermal vent,
at 250 to 300° under a pressure of 250 atmospheres. In the event,
this report proved to have been mistaken – it is not easy to do
bacteriology at 250 atm. pressure – and the record seems to be
held by a methane-producing bacterium called *Methanopyrus*, also

obtained from a submarine vent, which can grow at 112° under the necessary hydrostatic pressure.

All these 'hyperthermophiles', or 'ultrathermophiles' (scientists are not averse to a little drama in the names they give things), belong to a special group of bacteria called the Archaebacteria. This group is distinctive in many ways – I shall tell in another chapter how they represent a special class of living things – and they include both ordinary and thermophilic species. The great majority of hyperthermophiles are also anaerobes, which means organisms which live without oxygen. There is a simple reason why: the solubility of oxygen in such hot water is nearly zero, so there is little, if any, available. But hyperthermophiles still have waxy membranes, consist of proteins, and their genes are made of DNA; how do they manage to thrive in such heat?

Well, as far as their membranes are concerned, there are no ordinary fats that do not melt at 90 to 110°, and these organisms prove to have instead special waxy chemicals which are not true fats: they have higher melting points. (Digression for those who know organic chemistry: their fatty materials (lipids) are not esters but ethers.) Their proteins, too, prove to be naturally more heat-resistant than most, and there is evidence that some are permanently stabilised by chaperonins, others by newly-discovered substances that are not proteins, such as a compound of glycerol and phosphate which is plentiful in the record-breaking *Methanopyrus*. The problem of how such organisms manage their DNA is still unsolved, because their DNA is much like that of other living things: when extracted it unravels in the 65 to 75° range. It must be a matter of chaperonin-like chemicals, and there is indeed evidence that hyperthermophiles contain protein and protein-like materials which stick to DNA. The presumption is that these materials sustain the DNA in the double-helix state, despite the molecular battering caused by heat – except for that instant during cell multiplication when it must briefly unfold. But some of the other biochemical features of living things are not so tractable. As I wrote earlier, certain small molecules that are essential for energy transfer within cells, or for the processes involved in making use of genetic information, are simply decomposed by very hot water. Though some of these substances are chemically different, and more heat-stable, in hyperthermophiles, others are normal, and the upshot seems to be that these bacteria simply make them anew as fast as they decompose.

This is not so odd as it might at first sight seem. All living things, plants, animals and more ordinary microbes, are constantly turning over their cell constituents, renewing them as they break down or otherwise become ineffective. The impressive thing in the hyperthermophiles is the speed at which they do it.

Some microbiologists, reflecting on this particular aspect of thermophily, have predicted that the true upper temperature limit for life is around 160°, because a substance called ATP, which all known kinds of cell, including those of hyperthermophiles, use for energy transfer, breaks up as rapidly as it could be formed at such temperatures. They may be right – but a life form which uses something other than ATP is not unimaginable . . .

At the beginning of this chapter I asked why no plants and animals had learned to live at such high temperatures. In the light of what we know now, it was a loaded question. Consider what a plant or animal would have needed to learn. Every protein in its cells would have needed to become heat-resistant, or else super-chaperonins capable of protecting the sensitive ones would need to have been developed. Heat-tolerant waxes would have had to replace fats in cell membranes, in the sheaths of nerves and the like. DNA would have had to become heat-stabilised, probably by different chaperonins – and so on. Each one of these changes would have required a genetic change: a mutation in the organism's hereditary make-up. But during the evolution of higher organisms, mutations occur one at a time or, rarely, in small numbers. In higher organisms, evolution is a gradual, step-by-step process.

Some scientists have thought that there might exist a few, undiscovered, key genes which could cause massive subsidiary changes among the products of other genes, conferring heat-stability on a vast variety of gene products, without those genes having to mutate themselves. Evidence that something of the sort might be so has appeared occasionally in the microbiological literature. In 1932, a distinguished Dutch scientist, A. J. Kluyver, with his graduate student J. K. Baars, claimed to have taken strains of bacteria unable to multiply above 37° and to have 'trained' them to multiply at 55° by subjecting them to gradual increments of temperature. The evidence looked good; but the bacteria they used were of a kind which is particularly difficult to separate into single strains and, some three decades later, it became clear that they were mistaken: the populations they had started with were mixtures of ordinary and naturally thermophilic types living together. Nevertheless, strains of bacteria which ordinarily do not multiply at above 45°, say, can usually be trained in the manner of Kluyver and Baars so that their cultures grow up to about 50°. What happens seems to be that their whole temperature range is shifted upwards: their minimum growth temperature goes up, too. But the temperature increments are always a matter of a few degrees only.

In 1975, some excitement was generated among microbiologists by a claim in *Nature* that a common, moderate-temperature bacillus could be enabled to multiply at 70°, over 20° above its normal limit, by treatment with DNA extracted from a naturally thermophilic 'cousin'. Later the experiments were not successfully repeated, and the altered populations were lost, so the matter remained in doubt. A problem with bacilli is that, with some strains, the maximum temperature at which cultures will grow is much influenced by the culture medium: *Bacillus subtilis*, a very common species whose cultures grow well at 30 or 37°, normally will not multiply above 50°, but by judicious choice of medium they can be 'coaxed' to grow at 60°. However, that seems to be their normal limit and, in 1985, Mary Droffner and Nobuto Yamamoto of the Hahnemann University School of Medicine, Philadelphia, USA, reported experiments with such a strain of *B. subtilis* which went some way to confirm earlier findings. With some difficulty they obtained genetic mutants which would multiply at 68° – well above the normal limit. They then extracted DNA from one such mutant, treated an unchanged strain with it, and showed that its cultures acquired the ability to grow at 68°. All their strains had other distinctive genetic characters which enabled them to exclude the possibility of errors due to mixed strains or contamination in their cultures; they formed the opinion that two genes might be involved in conferring a modest level of thermophily on the moderate-temperature type.

It seems, therefore, that 'training', mutations and/or transformations can enable at least some moderate-temperature bacteria to multiply at up to 10° above their normal limit. However, the more we learn about the detailed requirements of hyperthermophilic life, the less likely does it seem that such changes could provide a step-by-step way in which the evolution of bacteria, let alone plants and animals, could lead to really substantial increases in heat tolerance.

One is compelled to the view that my question itself is misjudged. After all, some 3½ billion years ago, when the first primitive stirrings of life on Earth were taking place, the planet was hot. Of this there is little doubt. It is more than likely that, under a heavy atmosphere rich in nitrogen and carbon dioxide, the boiling point of water at the planet's surface was higher than it is today. So it is very tempting to imagine that the first autonomous living things on Earth came into being then, and were bacteria-like creatures, not just thermophiles but hyperthermophiles, capable of living in boiling or near-boiling water. I must come clean and say that there is not the slightest direct evidence for this belief, but the indirect case is compelling. As I have explained, it is difficult to imagine any way in which heat resistance could have been acquired during evolution. On the other hand, heat resistance could readily have been lost in a step-wise manner, as the planet cooled. One would imagine such loss to be followed by acquirement of membrane compositions, metabolic proteins and so on, which were better suited to the cooling environment.

In summary, gradual, step-by-step, loss of thermophily is easy to envisage; gradual step-wise acquirement of thermophily seems vanishingly improbable. So it follows that hyperthermophiles – early, pristine ones – were the common ancestors of today's living things, including ourselves. My question was indeed wrong: I ought to have asked, why it is that plants and animals have forgotten how to live at high temperatures? The answer then is easy: the planet had cooled long before plants and animals appeared, and their primitive ancestors had acclimatised to a cool world too. And by then thermophily, far from being an advantage, had become a restrictive nuisance, because a consequence of the machinery of the thermophilic cell being heat-resistant is that it works only slowly, if at all, in the cold. Areas of the planet where thermophiles might flourish were becoming fewer and fewer, and natural selection would favour the survival and persistence of organisms which were efficient at low temperatures. If this meant loss of heat tolerance, as it generally would, it did not matter.

But on a planet which is closer to its sun, especially on one with a heavy, Venus-like atmosphere, a hot, turbulent biosphere might have evolved. There might even be complex multicellular thermophiles, analogues of plants and animals. Life would be truly fast, as they consumed masses of nutrient to renew their rapidly decomposing metabolites, but provided water was normally liquid on their world, there is no reason why life should not persist and evolve. How the thermophilic equivalents of *Homo sapiens* would marvel at, and shudder at, those cold, slow-moving, slow-living inhabitants of planet Earth!

3

Cool, man, cool

Surviving cold

Humans are not hibernating mammals. They are obliged to keep warm and, if they get too cold, they die. Acute hypothermia is the name given to the situation in which this happens: people chill, become comatose and then die. Sometimes it occurs among incautiously adventurous people, who might go camping, swimming or hiking in dangerous weather. Most commonly it happens among old people who are too isolated, too poor, or too incapacitated to keep themselves adequately warm. It is a continuing social problem in countries with hard winters, such as Scandinavia and Russia; and in North America acute hypothermia still kills a few hundred people a year. Yet on rare occasions people have, seemingly miraculously, survived such chilling. In Norway in the 1950s, for example, a man was brought into hospital after 7 hours exposure to a sub-zero temperature (−1° Celsius; all temperatures in this chapter are in degrees Celsius). He was apparently lifeless and chilled so cold that a clinical thermometer did not register his temperature – which means it was below 25° (compared with our normal 37°), enough to finish most of us off in minutes. Yet after being gently warmed he recovered completely. Even more dramatic cases have been recorded, in which chilled patients have seemed lifeless yet have been revived, though most suffered severe frostbite. Because hospital facilities were available, such cases, although rare, are well-documented. A common feature seems to be that the survivors were usually drunk; not just ordinarily drunk, but boozed to a state of insensibility.

I am sure that it is this detail which lies at the root of a weird tale which I gleaned from a TV programme some 30 years ago. It told of a bizarre practice by a small rural community in a remote, mountainous part of New England, USA. Too poor to afford to feed and sustain their elders over the winter, we were told, the villagers would put them into cold storage. To do this, they would take them to a certain exposed mountain ledge – we were shown it, and bleak it was – and then get them paralytically drunk with whisky. Then they would leave them there to freeze, and to spend the winter covered in snow. Early in the spring they would return to the mountain, bring their elders home carefully, thaw and massage them gently, when they would revive, unharmed, for the summer.

The programme found locals to describe the process, but after this time I do not recall whether anyone appeared who claimed to have survived such crude cold-storage, or even to have actually participated in freezing a living relative.

Do not, please, believe this tale; it is medically impossible. The protective effect alcohol has on hypothermia, which is marginal at best, would vanish long before the human body approached freezing point, and this is no doubt why I have been unable to trace the tale again, either in medical texts or – more plausible sources – compendia of US folklore. But New Englanders, especially rural 'Yankees', have a reputation for 'putting on' strangers, and a quick-witted joker, picking up from newspaper reports a correlation between drunkenness and survival of acute chilling, could easily concoct a bogus folk-tale for a credulous television team.

If it was a New Englander's joke, it was a good one, because the involvement of whisky added just that touch of verisimilitude that a clever put-on should have. To a biologist, it gives the tale a fleeting touch of plausibility, because alcohol is an anti-freeze. It is not a good one, but it works.

By 'alcohol' I mean what scientists call ethanol, or ethyl alcohol, the component of wines, beer and spirits which makes people mellow or drunk as the case may be. But now I must stop using the word so loosely, because 'alcohol' is actually the technical name of a whole class of chemically related substances, most of which are rather poisonous. Indeed, as most people know, the ethanol we drink is a poison, but rather a weak one, and it is less toxic than most of the other alcohols; a healthy liver can detoxify modest amounts of it – given time. Most wines and spirits contain small amounts of more toxic alcohols as well, notably amyl alcohol (pentanol), which will give you a flush and a bit of a headache. Home-brews sometimes contain methyl alcohol (methanol, also known as wood spirit) which is dangerous: it will quickly make you ill and can blind and kill. Another alcohol sometimes present in traces is ethylene glycol (better-known as glycol, the motor car anti-freeze), which is quite poisonous and no fun at all, though some idiots have been known to drink it. Perhaps the most harmless alcohol is glycerol (its commercial name is glycerine), which is a perfectly normal component of some foods and fermented drinks. (The scientific names of alcohols, you will notice, all end in '-ol').

All alcohols are anti-freezes. Glycerol is an excellent anti-freeze. But it has another, more remarkable, property. It inhibits ice formation. This means that glycerol solutions remain fluid well below the freezing point of water, and when they get cold enough to solidify, the solid formed is not crystalline, like ice: it is vitreous, like glass.

Freezing kills mammals and reptiles because their cells become disrupted. But with isolated cells, as in blood or semen, a few survive. A lot of research has gone into finding out why most are

killed and a few are not, and the upshot is that ice formation is
the lethal event. Ice kills in two ways. On the one hand, it removes
water but leaves dissolved matter behind, which can become so
concentrated as to be lethal. On the other hand, shards of ice
disrupt the delicate system of membranes which protect the cell
and separate different zones in its interior. In consequence a
cell's contents leak and get mixed up, which kills it. But around
1948 Dr Audrey Smith and her colleagues of the Medical
Research Council made a most important discovery: if some
glycerol was added to semen, the sperm were protected from
freezing damage.

Incidentally, their discovery involved the sort of accident which
has not uncommonly led to scientific advance. They were testing
chemicals for protective effects on sperm, and they found that a
solution labelled 'laevulose' (a kind of sugar) prevented killing
even after cooling to −79°. But the bottle proved to be mis-
labelled: it actually contained 10% glycerol plus 1% of a protein
called albumin. It ought also to have contained a disinfectant
called thymol, but this had been left out – a second happy mistake,
because thymol would have killed the sperm at once. The matter
was sorted out, the protective effect of glycerol confirmed, and
the post-war era of artificial insemination and organ storage was
ushered in. For their finding led to the discovery that glycerol,
and many other substances too, would protect all sorts of cells,
and even whole organs, from freezing damage, and permit their
cold-storage for transplantation, implantation, transfusion, artificial insemination and all such marvels of modern human and
veterinary medicine.

The paper which Dr Smith and her colleagues published on
the subject in *Nature* in 1949 not unreasonably made no mention
of their fortunate mishaps with laevulose.

Why does glycerol protect? Because it gets into the cells, as
well as surrounding them, and prevents ice formation. It is as
simple as that. If the temperature goes low enough, a non-
crystalline glass is formed. A wide variety of other substances act
like glycerol in this respect. Scientists call them 'cryoprotectants';
as well as alcohols, they include sugars, solvents, non-toxic detergents and even proteins.

Curiously, some insects have 'known' this for millions of years. There are numerous kinds of arthropod which survive freezing – as was first reported by the French naturalist Réaumur in 1734 – and can therefore inhabit tundra regions, and Arctic and Antarctic zones. They respond to temperatures below freezing point by generating anti-freezes in and around their cells – often glycerol, but sometimes other alcohols (including, in a certain bark beetle, ethylene glycol). Frost-tolerant plant cells, too, have proteins which act as anti-freezes.

Bacteria (except when they have formed spores) are as susceptible as mammals to the lethal effects of freezing. Again, glycerol works very well with bacteria and freezing cultures to which glycerol has been added is a time-honoured way of preserving precious strains of bacteria in the laboratory. The best environment for this is a solution containing 10 to 30% of glycerol, and it does not matter if other materials, such as salts and nutrients, are present as well. Other cryoprotectants work equally well; they are effective at a variety of concentrations and the strong solutions used with glycerol are not always necessary, but few are as reliable for long-term storage as glycerol. And to go back to those New Englanders for a moment, ethanol, when it does not kill bacteria, can be a cryoprotectant. Could it cryoprotect a drunk? Never!

Let me return to bacteria. An important consequence of cryoprotection is this. When bacteria are killed by freezing, without a cryoprotectant, most of the cells burst open, and this releases their contents into the environment. Among those contents are proteins and sugars which can themselves be cryoprotectants. Therefore, provided the population is dense enough for the natural cryoprotectants to reach an effective concentration, the first ones to die protect a few of their neighbours. Put another way, though individual bacteria are as likely as any organism to be killed by freezing, their biochemistry is such that dense populations can become cryoprotected. This is very important as far as the survival of bacterial species in nature is concerned: in a protracted and severe frost, for example, such as would kill most higher organisms, many individual bacteria will die too, but the species will survive.

The cool life

But survival, in what amounts to a dormant state (like a deciduous tree in winter), is hardly living, even to a microbe, whose sole objective in life (if it were capable of being aware of such a thing) is to increase the total number of replicas of itself. (Not much different from rabbits, you say? Or people? No, not much. Save that point for another time!) Even among cold-blooded animals, and most plants, vital processes tend to slow down to a near standstill when the temperature gets below 10°. Only the few specialised plants and the invertebrates which I mentioned earlier can struggle on at temperatures closer to freezing. But biologists have been aware since the 1880s that microbes exist which live out their lives near the freezing point of water; in really cold places microbes, especially bacteria, are present in great variety.

The world, though few people realise it, is really a very cold place, and not just at the polar regions. The warmer part consists of much of the land surface, plus a mile or so of atmosphere and a few tens of metres of the oceans' surface; these are the only parts which experience the season-related temperature fluctuations with which we and other higher organisms are familiar, and which keep our environment within the equable range of 10° to some 45°. But more than two thirds of the Earth's surface is covered by sea, and two thirds of that is unaffected by changes in surface temperature and stays close to 2°. Looking at the matter another way, four fifths of the inhabitable volume of this planet (in which I include the polar ice-caps) never rises above 5°. It is a chilly world, and most higher organisms have colonised only the warmer bits; the rest is home for a myriad of microbes.

Microbiologists distinguish two types of cold-tolerating microbes. 'Psychrotrophs' are those which, while tolerant of cold, prefer warmth; 'psychrophiles' are wholly adapted to cold and die at temperatures above 20°. Those who go skiing might have encountered 'pink snow', a bloom of psychrophilic algae which sometimes appears on snow in the Swiss Alps: it is green at first but the sun bleaches the chlorophyll and it turns pink. That particular alga grows best at 2°. An important habitat for comparable microbes, mostly bacteria, is the bottom of the sea, and there they live on the sediment that falls from above: organic matter sinking from the surface plankton, the excreta of fish, particles carried from the land by wind and birds, and, increasingly, muck discharged by people from ships. The hydrostatic pressures are high, especially in the really deep ocean trenches which go down a couple of miles, but the bacteria do not mind at all. Apart from the deep sea, psychrophiles and psychrotrophs have other homes: arctic ponds, deep alpine lakes, and tundra soil near the perma-frost region, all of which support abundant microfloras. There are a few places, such as brackish pools in Antarctica, where the water remains fluid even below $-5°$ because dissolved salt pre-vents freezing, yet microbial growth and activity goes on there: the lower temperature limit for such activity seems to be about $-12°$. And closer to home, mankind has provided such bacteria with a new habitat: domestic refrigerators are usually set at 4 to $6°$, and in these cheese can go mouldy, or orange juice can become yeasty. The majority of such cold-tolerant microbes are harmless to people, but a few cold-tolerant bacteria exist which are capable of generating tummy upsets and may contaminate refrigerated food. They present a problem of which the food industry, which depends greatly on chilling for short-term food preservation, is only too well aware.

Psychrophilic and psychrotrophic microbes have to solve a special biochemical problem. It is an axiom of chemistry that the cooler the environment, the more slowly do chemical reactions take place. As far as the chemistry of living things is concerned (in which enzymes promote all sorts of otherwise unlikely chemical reactions) this axiom is abundantly true: many biochemical pro-cesses slow up disproportionately as the temperature declines, and may grate to a virtual halt around 10 to $15°$.

But 'virtual' is the crucial word here: most of them do go on, if painfully slowly. Correspondingly, the processes leading to the death of cells slow up too. The psychrotrophs take advantage of this situation: essentially they are ordinary bacteria that can live out their lives, albeit very slowly, at low temperatures. The major-ity of common refrigerator contaminants are of this kind: ordinary microbes which like it warm but which are just ticking over in the cold. Psychrophiles actively prefer low temperatures, and grow in the cold almost as fast as ordinary bacteria do in the warm. It transpires that they often possess special kinds of enzymes which actually work faster at low temperatures than the corresponding ones in normal bacteria, but they have a penalty to pay: these hyped-up enzymes are more easily destroyed by heat, which may explain why such psychrophiles die should temperatures rise much above about 20°.

Some of the enzymes of psychrophilic bacteria can be extracted and purified; their ability to work well at low temperatures has proved attractive to biotechnologists because they can be exploited industrially, for low-temperature laundering or waste treatment, for example. But from the microbes' point of view, they serve to keep the cell's biochemical machinery going in the cold. Yet this is the least of a psychrophile's problems. A more serious one is keeping its own surface from setting hard. Bacteria have two skins, an outer one which is a stiff molecular mesh, through which molecules of food and water can diffuse fairly easily, and an inner one, elastic and membranous, which has to be very selectively permeable, so that nutrients can get in but the internal substances of the cell do not leak out. (This, by the way, is the skin which ice damages lethally; the outer layer is tougher and serves to keep out big molecules and to sustain the cell's shape). The cell membrane, as it is called, includes a lot of fat in its structure, and its permeability is very much influenced by the fluidity of that fat. Anyone who has stored butter in the refrigerator knows what happens to fat at low temperatures: it goes hard. Once that happens to a cell membrane, transport of food and water into the cell, and of waste products out of the cell, slows up and may stop. Psychrophiles have cell membranes of a special fatty composition, such that they are relatively fluid at temperatures near freezing point – and again they pay a price: their membranes become too fluid, and begin to melt, when the environment warms to the temperatures that most bacteria prefer.

This fact explains a paradox that was resolved only around the middle of this century. For some decades the real existence of psychrophiles was doubted because, though there seemed to be evidence for them in nature, the cultures of bacteria from cold places isolated in the laboratory almost always proved to be psychrotrophs: tolerant of cold but happiest in the warm. In the end the paradox proved to have arisen from a technical error: samples taken from cold places, which were often remote, warmed up on their way to the laboratory, so the psychrophiles died and only the psychrotrophs remained. Once this fault was realised, the fact that true psychrophiles are abundant in cold places became clear.

Actually, most bacteria, including psychrophiles, can, given time, adjust within limits the sorts of fat composing their membranes to acclimatise to an altered temperature, be it upwards or downwards. Psychrotrophs do just that: they alter their membrane composition to acclimatise themselves to cooler environments, and they need time to do it: they must cool slowly.

Psychrophilic microbes, and psychrotrophs too, include representatives of a wide variety of microbial genera; most psychrophiles are bacteria, but psychrotrophs include fungi and micro-algae in addition. In their chilly habitats, these microbes conduct all the processes that make the warmer, everyday environment tick over: organic matter and detritus are decomposed, the essential biological elements are recycled, and, where sunlight reaches, psychrotrophic photosynthesisers make new organic matter from carbon dioxide. In those cold, remote and often dark places there is a teeming cosmos of living things recycling the biological elements, and most higher organisms are wholly unaware of it. And a Good Thing that microbial cosmos is, too. For if it did not exist, over the aeons the oceans would have become the dustbin of the non-psychrophilic, non-psychrotrophic world, crudded up with organic matter, with its carbon and accompanying biological elements locked away for ever. The natural cycles of the elements which sustain us today would have been drastically depleted. No-one could guess what the living world would then have been like, but I doubt very much whether humans would have been here to think about it all.

4

The big squeeze

In the last two chapters I wrote about life at extreme temperatures: how certain types of bacteria live at temperatures at, and even above, the ordinary boiling point of water, and cannot manage at ordinary temperatures; and how other types exist which flourish in excruciatingly cold places and find our ordinary world too hot. You hardly need to be a microbiologist to know that temperature is a very important determinant of the sorts of creatures which flourish or die out in various localities, but bacteria have colonised hotter and colder environments than have any other forms of life.

A different physical factor that determines what sort of creature lives where is pressure. Yet though many people have barometers in their houses, few regard pressure as determining anything but the weather – and that, too late to be of much use.

Let me digress on the importance of pressure in our lives. Have you, for instance, ever worked out what weight of air there is in your sitting room? It can be done quite easily. The arithmetic is based on an axiom of chemistry which says that, at similar temperatures, all gas molecules occupy the same volume, whether the molecules are large or small. The reason is that gas molecules, like the molecules of solids and liquids, are in constant motion, but gas molecules are specially energetic: they bounce around so vigorously that the space they take up, bumping their neighbours aside, is very much bigger than they are themselves. So their own volume makes no real difference to the volume they effectively occupy. Like other 'gas laws', as they are called by chemists, this one is not quite true: the actual volume of a gas molecule *does* make a minor contribution to the volume it occupies, but for present purposes this does not matter.

Scientists express the weights of molecules in units based on the weight of a hydrogen atom. This weight is taken as 1 unit. The reason why they do not use ordinary weight units, such as grammes, is simply that the resulting numbers would be idiotically small: a hydrogen atom weighs one sixth of a hundred thousand billion billionth ($^1/_6 \times 10^{23}$) of a gramme, and it is the lightest atom of all (which is why its weight has been chosen as the primary unit). Air is a mixture of four parts of nitrogen to one of oxygen, plus tiny amounts of other gases (which can be disregarded for now). Molecules of nitrogen, which actually consist of two nitrogen atoms stuck together, are twenty-eight times as heavy as atoms of hydrogen, so they weigh 28 units each. Molecules of oxygen, which are also pairs of atoms, weigh 32 units each. Hydrogen molecules, incidentally, also consist of two atoms, so they weigh 2 units – having the lightest of atoms, hydrogen is also the lightest gas of all.

It follows from the axiom I mentioned that, if you take an amount of gas in grammes equal to its molecular weight – 28 grammes of nitrogen, for example, or 2 grammes of hydrogen – it will occupy the same volume no matter what gas it is. Chemists confirmed this in the mid nineteenth century and observed that the volume of the molecular weight in grammes of a gas is 22.4 litres at ordinary temperatures and pressures. Chemistry students have to learn that number early in their training.

Knowing that number we can do our sum. The molecules

composing air have an average molecular weight of 28.8, so every 22.4 litres of air in your sitting room weighs about 28.8 grammes. A suburban sitting room of moderate size, say 4 × 5 × 3 metres, has a volume of about 60 cubic metres. There are 1000 litres in a cubic metre, making 60,000 litres. Divide that volume by 22.4 and multiply by 28.8 and there you are: there are 77,143 grammes of air in your room. That is 77.74 kilogrammes, or about 170 lb. of air: equivalent to the weight of a healthy 6 ft. man.

Air is heavier than many people think – which is why air pressure can be so destructive when there are gales around. When we walk across a room, we push many kilogrammes of air around. And everywhere, indoors or outside, a column of atmosphere presses down on us at a pressure of 14 lb. per square inch (1 kg per square centimetre).

But we are used to it. We have barely any sense of air pressure and, except when the wind blows, we are hardly conscious of our surrounding air at all.

People can sense changes in pressure. Air travellers learn this, sometimes uncomfortably: when an aircraft ascends, or especially when it descends, some passengers feel discomfort in their ears as the air pressures on either side of the eardrum equalise. If one has recently had a cold, the experience can be really painful; sucking a sweet often helps to stop it happening by making one swallow, a movement which helps to equalise the pressure across the eardrum. But this is a transient response to changing pressure; we do not sense air pressure if it remains steady, or even if it changes slowly. People can adjust to lowered air pressure provided they still get enough oxygen: mountaineers ascending really high mountains, where the atmosphere is thin because the pressure is low, wear oxygen masks, devices which top up their oxygen supply; and early space vehicles had atmospheres of almost pure oxygen, set at one fifth of the normal atmospheric pressure. That pressure, one fifth of atmospheric, is about the lowest that humans can comfortably manage with, simply because they would become oxygen-starved if the pressure were much lower. Similar limits probably apply to most terrestrial living things, plant and animal, though they would never encounter such low pressures naturally. Some kinds of mudworm, and certain air-breathing microbes, would be expected to cope with lower pressures because they are especially good at picking up oxygen even when it is in short supply. Bacteria of the class called anaerobes – a class which is able to grow without oxygen – can tolerate much lower pressures: they grow happily in as near a vacuum as is compatible with having water around (their habitat is water so there has to be some about). Such bacteria do sometimes encounter low pressures: they can be a nuisance in vacuum-packed foods which, despite the description, do not contain a true vacuum, just a low pressure of water vapour.

Bacteria, then, beat other living things in coping with the lowest end of the pressure scale. What about elevated pressures? As far as land creatures are concerned even the highest atmospheric pressures present no problems: in the deepest coal mines at some 1500 metres (4900 ft.), the air pressure is only about a sixth greater than at the surface. But in water things are very different, and water – rivers, lakes and seas – is an important habitat for living things.

Hydrostatic pressure increases by an amount equivalent to one atmosphere for every 10 metres depth. (This is why divers cannot descend more than a few metres without special suits, or in submersibles of some kind: such high pressures entrain several kinds of physiological damage even if the diver is protected from drowning.) Even in coastal waters the bottom of the sea may be at several atmospheres pressure, and at the bottom of the deepest parts of the ocean, in what are known as 'trenches', which can be some 11 kilometres deep, the hydrostatic pressure is equivalent to 1100 atmospheres.

Ordinary fish cannot survive even at much more modest depths. If they ever got down that deep, they would, in a sense, be squashed. Instead, specially adapted kinds of deep-sea fish have evolved, as well as squids and other marine creatures, which live out their lives at such pressures, and they die when brought to surface waters. And down on the floors of the deep trenches there are animals: large tubular worms, a few kinds of shellfish, specialised crabs. They tend to cluster around centres of geothermal activity, especially where warm water emerges from the ocean bed, and for a rather special reason. It is here that microbes flourish and provide the basic food on which the other organisms subsist (I wrote about the kinds of microbe in Chapter 2). They obtain their food from the emerging hydrothermal waters and multiply, and the more complex organisms browse on them. Some of the higher organisms, the worms and clams, have populations of bacteria living in their guts or gills, collecting and supplying food for the host animal.

Elsewhere on the deep sea bed there seems to be no life but for a few bacteria, living and partly living on detritus falling from the upper layers of the ocean, detritus which is almost devoid of food value, having been gone over again and again by other microbes on its way down. These specks of life live and metabolise very slowly, taking months to multiply, and they are exceptionally resistant to starvation. They have a hard life, if one can regard bacteria as having a quality of life at all. The presence of live bacteria at these depths in so harsh an environment might seem surprising – almost nothing to eat, squeezed at some 15,400 lb. per square inch, and icy cold at 2.5° Celsius year in, year out. But to a microbiologist it is not surprising. Bacteria are single, undifferentiated cells and, unlike more complex creatures, they possess no organs comparable to eyes, guts or swim bladders which might squash at high pressures or, if adapted to high pressure, might swell and disrupt in normal conditions. Pressure ought not to matter to bacteria, and indeed, ordinary surface bacteria, such as *Escherichia coli* from human intestines, have proved able to grow in the laboratory at 300 atmospheres, given appropriate nutrients.

Therefore, unlike deep-sea fish, deep-sea bacteria ought not to be incommoded when they are brought to the surface. There has been a lot of argument among microbiologists about whether they are or are not. For one thing, it is extremely difficult to do laboratory experiments with bacteria at such high pressures: reinforced steel equipment and the risk of explosion severely restrict one's scope! Yet some marine microbiologists have managed to work in this area, and for several decades in the middle of this century it seemed that many deep-sea bacteria did require high pressures to grow. Such bacteria were given the name 'obligate barophiles', meaning that they were obliged to live under pressure. But some of the experimental difficulties had confused the earlier researchers, and more recent work has not confirmed this belief as far as free-living deep-sea bacteria are concerned. They all seem able to grow at atmospheric pressure provided the temperature is low and the nutrients are right. However, genuine obligate barophiles have been discovered in one remarkable circumstance. Several types have been found inhabiting the intestines of deep-sea animals such as fish, arthropods and worms. One such organism, a spiral-shaped bacterium, grows best at 500 atmospheres pressure in cold water at 2 to 4° Celsius! That organism can tolerate atmospheric pressures, but grows very slowly; another type which also enjoys such high pressures actually dies in a few hours at ordinary pressures.

Bacteria which have existed for billions of generations at atmospheric pressures seem to be more or less indifferent to a 300-fold increase in pressure; and bacteria which have acclimatised themselves to monstrous hydrostatic pressures for goodness knows how many generations do not mind a 500-fold decrease in pressure. So why should obligate barophiles exist among bacteria at all? There seems to be no good reason, and the fact that the few known examples all live inside higher organisms makes the question even more perplexing. Perhaps further study of those examples will give scientists a clue, though progress is unlikely to be rapid, given the difficulty of doing research at high pressures.

But there is a broader message in all this: pressure may be a critical matter to higher animals and plants, including ourselves, but neither the highest nor the lowest pressures encountered in the habitable parts of this planet's surface are obstacles to the establishment of microbial life. Therefore pressure is no obstacle to the development of more elaborate living things, dependent on those microbes for their nutrition and life cycles. Low pressures are too rare on Earth for such communities to have evolved here, and our high-pressure communities are restricted to the ocean deeps. But the Earth might have been virtually airless, as Mars is today, or it might have been massive, like Jupiter, with a gravity which leads to huge surface pressures. In either case, given otherwise equable conditions, life of the terrestrial kind could still have appeared, and evolved into complex beings. They would have been very different from living things as we know them today, but their basic biochemistry could well have had a lot in common with that of our present microbes.

5

A salty tale

Salt and water

Consider a salt solution, seawater perhaps, drying out. As the water evaporates, the solution becomes more and more concentrated until it reaches saturation; then crystals of salt begin to separate out. Conversely, if one sets about dissolving salt in water, the solution will become more and more concentrated until it saturates: no more will dissolve and salt crystals will remain at the bottom of the vessel. Everyone who learns about this sort of thing is taught to regard the salt solution as becoming more and more concentrated – and if one adds extra water at any stage, the solution becomes more dilute again.

But there is another way of looking at the whole affair. The more salt one adds, the more it dilutes the water. This is not mere word-play; it is a reality which is responsible for most of the biological effects of salt. To explain why, I must describe what happens when salts dissolve in water.

Common salt is sodium chloride, which means that it consists of atoms of sodium (a metal when it is not in chemical combination), and of chlorine (a gas when uncombined) in equal numbers. These atoms are held together by a strong electrical interaction, the sodium atom having a substantial positive charge and the chlorine an equal negative one. Notice that I have avoided the word 'molecule'. Though the smallest unit of sodium chloride would be one atom of sodium bound to one of chlorine, and the ratio of the two atoms is always 1:1, in fact the units join up in three-dimensions to form a lattice: a single crystal of sodium chloride can be looked upon as being a giant molecule. Water, in contrast, is made up of real molecules – two hydrogen atoms stuck to one oxygen atom – and the component atoms have only slight charges: oxygen a bit negative, hydrogen a bit positive. This property, which is called polarity, nevertheless causes water molecules to tend to associate in chains or clusters, because the attraction between a positive pole on one molecule and a negative pole on another pulls them together momentarily. But the electrostatic forces between the molecules are so weak that the structures are constantly disrupted by thermal motion: they form and break down in a continuous kaleidoscope of movement.

When common salt dissolves in water, the electrical charges on the sodium and chlorine attract water molecules quite strongly, and they cluster round both. This enables the sodium and chlorine atoms to separate and dissolve as charged particles. Chemists call such particles ions. (Chemists often talk of ions simply as charged atoms. The sodium and chlorine atoms in a crystal are actually ions, but in watery solutions ions are always surrounded by a cloud of water molecules, held by electrostatic attraction.) When water molecules are stuck round an ion, they cannot move freely: some may be bumped out of the cloud by the thermal movements of neighbouring water molecules, but they will be replaced almost instantly. In effect, the more ions there are – which is to say, the stronger a salt solution becomes – the lower will be the proportion of water molecules which can move about freely.

Common salt is far from being the only substance which dissolves in water as ions; many other salts, such as sodium bicarbonate, magnesium sulphate or potassium nitrate, do the same thing. They all 'dilute' the water disproportionately by entrapping water molecules in a cluster round their ions.

The same principle, that dissolved matter dilutes water, is true of all water-soluble substances, even if they do not form ions. Sugar molecules do not form ions in water; in their case they simply dilute the water by being very large. A molecule of cane sugar is nineteen times the size of a water molecule so, in a strong solution, sugar displaces a lot of water.

Living with a pinch of salt

But let me return to common salt. The way in which it interacts with water matters very much to living things, whether they be plants, animals or microbes. The reason is that cells of living things, those actively conducting normal life processes, rely on water molecules being able to move into and out of them freely. (I specify active cells because dormant cells, such as those comprising spores, seeds and other bodies which can resist drying, become almost impermeable to water.) Water flows in and out of cells easily, though the substances which may be dissolved in it generally do not; sugars and salts, for example, only get into cells if there are special mechanisms for letting them in – as there usually are. Cells find a through-put of water essential both for the uptake of nutrients and for the excretion of wastes. On a gross scale, it is the reason why we need to drink and urinate, why plants need watering, and why microbes have to have a watery habitat – even if it amounts only to the minuscule film of water on skin, on a soil particle, or a leaf surface.

It is a fact that all terrestrial life keeps itself wet; where living things have truly dry bits, they are dormant (e.g. seeds or spores), or dead (e.g. the carapaces of insects, old wood, human hair or finger nails).

Cells have to control their internal compositions carefully, and for this reason the living protoplasm is enclosed in a special kind of membrane, one which has the important property of letting water in and out freely, but of excluding or regulating the passage of most dissolved substances. Membranes of this kind are called 'semi-permeable', because they are permeable only to certain substances, and have been known to scientists for several centuries: for example, the linings of animal intestines have long been known to be semi-permeable. If you place a sugar solution on one side of a semi-permeable membrane, and pure water on the other, water molecules move freely through the membrane in both directions, simply because water molecules are bustling around in a state of constant movement. But more water molecules are going into the solution than are coming out, simply because many of the molecules on the solution side are sugar and bounce back, without getting through. The upshot is a net flow of water into the solution. If the two sides are enclosed, the water pressure builds up on the solution side until it is sufficent to squeeze water back through the membrane as fast as it comes in.

Salts as well as sugars, indeed all dissolved substances, generate pressures of this kind across semi-permeable membranes. Chemists call the water pressure which a solution can generate its 'osmotic pressure'; I shall need to use that term occasionally in this chapter.

Cells, as I said, enclose their protoplasm in a semi-permeable membrane. Protoplasm has an osmotic pressure. It is a solution of proteins and minerals dissolved in water, in which more elaborate organelles and particles float. The minerals, which are essential to life, are mainly salts of magnesium, potassium, sodium and calcium, mostly as positive ions, of course, and mostly balanced by negative chloride, bicarbonate and phosphate ions. Cells regulate their content of such salts very precisely, because if the salts get out of balance the cell machinery does not work properly: the total salinity of protoplasm, the net concentration of all its salts added together, is about a third that of sea water (roughly equivalent to a 0.85% solution of sodium chloride). By manipulating the concentrations of other substances dissolved in their protoplasm, cells also regulate their internal osmotic pressures, and they keep them slightly above that outside. This causes water to try to flow inwards, keeping the cell turgid, like a pumped-up tyre, which is a healthy condition for all types of cell to be in.

If a cell finds itself in a strong, say a saturated, salt solution, things go badly wrong. Its membrane can, and does, keep out the ions comprising the salt. But the water molecules in the salt solution are clustered round ions and can barely move around; instead of flowing in, few will permeate a cell. As far as a cell is concerned, that solution seems virtually dry. Moreover, like any dry environment, but because of its high osmotic pressure, it tends to draw out, to soak up, water from within the cell: effectively it dehydrates the cell.

A salt solution need not be saturated to dehydrate a cell: it need only be strong enough to generate a higher osmotic pressure than the cell can muster within itself.

Under a decent microscope one can watch what happens when a strong salt solution dries out a cell. If you take cells from a crushed plant root, for example, and look at them in water, you will see the textbook spectacle: a rigid cell wall enclosing protoplasm – the protoplasmic membrane will be invisible, pressed tightly against the cell wall. The protoplasm will seem to flow gently, carrying tiny particles about, and will also contain a nucleus and some other granular objects: perhaps, too, there will be a space – the vacuole – which seems to have nothing in it (but actually contains sap). Expose the cells to a moderately strong solution of common salt, and the protoplasm, with its vacuole, can be seen to shrink, eventually contracting to a residual blob of jelly. This happens with animal and most microbial cells, too.

Most cells can recover if such a stress is very brief, but they very soon die if the stress is prolonged. Solutions of other salts, such as sodium sulphate or magnesium chloride, have a similar effect, and so do solutions of non-permeating, non-ionic substances such as sugar.

Strong brines or strong sugar solutions, then, kill cells by dehydration and, naturally, prevent their growth. This is why salting, or preserving in sugar, have become traditional ways of preserving foods from the ravages of noxious microbes and moulds (maggots, too, since their eggs are also dried out). It is also why the Dead Sea got its name: it is a huge inland lake that has become so salty that neither vegetation, nor fish and other marine animals, can live in it: it is a 24% solution of salts of various kinds, nearly ten times as salty as seawater.

The sea represents more or less the upper limit of salt tolerance for the great majority of living things. At slightly under 3% salinity, most of the dissolved matter being common salt, it is just too salty for most land and freshwater creatures. Few higher plants can stand its salinity, but seaweeds, which are large algae, flourish there and, of course, fish and a vast menagerie of marine animals including arthropods, molluscs, worms, corals, and even marine mammals. A number of these creatures can tolerate both sea and fresh water, and can move freely from one to the other. Then there are numerous types of specialised marine bacteria, some of which live in either fresh or sea water; and others so specialised that they can grow only in sea water. In between are types that prefer either sea or fresh water but can acclimatise to either.

The ways in which marine creatures adjust to the saltiness of the sea are various, and are not strictly relevant here, because I am concerned with life in more extreme conditions. In essence, they survive by keeping the salt out of their systems and keeping water in. Bacteria have a special problem, being tiny single cells, but most respond to the saltiness of their environment by adjusting the amount of dissolved matter in their protoplasm to keep it in balance with the outside. For example, *Escherichia coli*, a bacterium accustomed to the low-salt conditions inside our intestines, responds to a mild salt stress by making an ionic substance called betaine within itself. Salt dilutes water, so a salt solution tends to draw water out from inside cells; *E. coli* turns the tables by diluting its own water with betaine and reversing that tendency. A good idea, but in *E. coli*'s case it does not always work all that well: sea water contains too much salt for it to adjust to, and kills it after a couple of hours. Happily, sea water does that to most pathogenic intestinal bacteria, too, or our polluted beaches would be more unhealthy than they are already.

Living with a lot of salt

The sea is a habitat for a myriad of living organisms which have adjusted to its salinity in their various ways. The Dead Sea is one example of a number of places around the world which are so salty that even marine creatures cannot live there. Others include the Great Salt Lake in Utah and numerous coastal regions where land-locked sea water has evaporated and become concentrated – either for natural geographical reasons or because people arranged it deliberately for the manufacture of sea salt. Natural environments of this kind have existed for millions of years; mineral salt deposits were formed by evaporation of ancient seas or land-locked lakes during geological time. Higher organisms have never learned to colonise them but, I need hardly say, microbes have. And since they have possessed this habitat on their own for countless millenia, some prove to be very peculiar biologically.

If you fly into San Francisco International Airport, coming in from the West over the Bay, you will see huge salt pans ('salterns' is their correct name) close to the shore, where sea water is evaporating and sea salt is being harvested. And you will notice a peculiar thing about them: they are not white, as salt ought to be, they are pink, or pink and white in patches. The reason is that they are colonised by salt-tolerant microbes which are red, and the ones responsible for most of the colour are highly specialised bacteria called halobacteria ('halo-' from the Greek *halos* meaning salt; they are not sanctified). Also contributing to the pink colour is an alga called *Dunaliella*, though this can be green in less salty conditions.

I shall say something about *Dunaliella* first. Algae are altogether more complicated organisms than bacteria, but as algae go this one is structurally rather simple: a microscopically small, single-celled creature, not a great frond-forming seaweed. Algae are primitive plants; they use sunlight to make starch from carbon dioxide and water – the process called photosynthesis – and they make the rest of themselves from the starch plus a few minerals. For photosynthesis, plants need a green pigment called chlorophyll – which is why leaves are green – and *Dunaliella* is no exception. But the green colour can be masked by large amounts of a red pigment called carotene. This substance is present in green plants as well, though generally in much smaller amounts. (It also masks chlorophyll in copper-leaved varieties.) Chlorophyll is destroyed by excessive sunlight and carotene is there because it protects chlorophyll from light damage. *Dunaliella* certainly has need of such protection because, when it is growing on the surface of a saltern which is being evaporated by a strong sub-tropical sun, the illumination is harsh and intense. (Its environment may also get rather hot, but it can tolerate up to about 40° Celsius.) However, *Dunaliella* seems to have much more carotene than is needed just for protecting chlorophyll: carotene can amount to a fifth of the cell's dry weight. Why it has so much is still a puzzle, but carotene is a valuable food supplement (the human body makes it into vitamin A), so *Dunaliella* is used commercially as a source of carotene.

But I digress. Among microbiologists *Dunaliella* is famous for its wide range of salt tolerance: from about 2% common salt (more dilute than sea water) to saturation, which is 33%. Actually, this is not a record, for two reasons. The first is that about half a dozen species of *Dunaliella* are known, and no one species can cope with the whole range; the second is that a bacterium called *Halomonas elongata* walks off with the prize by being able to grow with anything from 0.3 to 33% salt. How it does this has not yet been worked out.

Dunaliella, however, has been studied extensively, and to cope with salt it adopts something like the strategy I mentioned with regard to *E. coli*: it makes a substance inside itself which counteracts the dehydrating effect of salinity. On sensing increased concentrations of salt – and how it senses that is a whole new subject which I cannot go into now – *Dunaliella* immediately starts getting rid of any sodium ions which might have leaked into itself and replacing them by potassium ions. These do not go on accumulating, however; their increase acts as a signal within the cell, which tells it to alter the pathway of its photosynthesis so that, instead of making starch, it now makes glycerol. Now, glycerol is a water-soluble, non-ionic molecule which, in *Dunaliella*, counterbalances the dehydrating effect of the external salt, just as betaine does for *E. coli* (which, incidentally, also uses an increase in potassium ions as an early warning). The amount of glycerol which *Dunaliella* can accumulate is again remarkable: up to 30% of its dry weight. (What with that and 20% carotene, *Dunaliella* can accommodate a lot of biochemical baggage!) Again, this high glycerol production has been exploited commercially.

All the glycerol accumulated by *Dunaliella* certainly 'dilutes' the internal water, but glycerol is one of the most harmless biochemicals known. It is a liquid itself, it mixes completely with water, and water molecules remain reasonably mobile when mixed in with glycerol; in effect, *Dunaliella* makes do with a somewhat dehydrated protoplasm in which all its biochemical machinery works adequately. But the glycerol seems to be something of a disadvantage because, should the external salinity decrease, *Dunaliella* gets rid of as much of its glycerol as it can, excreting some and making the rest into starch.

Dunaliella has discovered an ideal substance with which to protect its innards from salt damage and dehydration. Why does only *Dunaliella* do this? Well, there are one or two other genera of micro-algae which mimic *Dunaliella* in that way, but there is a fundamental reason why the majority of microbes cannot exploit glycerol. Quite simply, they cannot hold on to it. Glycerol is a remarkable substance biologically in that, as far as the vast majority of cells is concerned, it permeates just as readily as water: it flows freely in and out. (In Chapter 3 I told how this property is exploited by biologists to protect bacterial cultures from death by freezing, and to deep-freeze living organs for transplants.) In effect, the membranes surrounding the protoplasm of bacteria, and the cell membranes of most other creatures, cannot tell the difference between water and glycerol; they let both in and out freely. *Dunaliella* has developed a most unusual kind of membrane that enables it to retain glycerol, and to excrete it in a regulated manner if need be. It is this property which has enabled it to colonise its curiously unpromising habitat.

Dunaliella is but one inhabitant of strongly saline places. As I indicated earlier, several types of bacteria live there, too. Some look curious: the only known examples of flat, square bacteria have been found growing on salt crusts in salterns and natural evaporating pools. They have not been cultured successfully in a laboratory. Perhaps the most fascinating are the red halobacteria I mentioned earlier, species of a genus called *Halobacterium*. They are, incidentally, responsible for red stains which appear when salt-preserved food, such as salt fish, goes 'off'. Under the microscope their appearance is rather ordinary: they are rods, usually short and sometimes with box-like ends. It is their way of life which is their curious feature. For one thing, they absolutely require salt in their environment: they need at least 18% sodium chloride to grow at their best, and flourish in saturated salt solutions and on damp salt crystals. And unlike all other microbes, they do not exclude salt from their interiors: they admit salts, albeit selectively. If they have grown with, for example, some 30% of sodium chloride, they do indeed keep a lot of it out, but even so their protoplasm contains as much as 5%. However, their protoplasm is awash with potassium chloride, which can be as strong as 30%. Salt concentrations of this magnitude, whether they be potassium or sodium salts, would have a devastating effect on the proteins of ordinary bacteria. Ordinary protein molecules would become distorted; they would clot and fall out of solution, rather like beaten egg white; their biological activities would be destroyed. However, the proteins of halobacteria are special; they do not react to salts in this way. Moreover, most cell proteins are enzymes, performing the multiplicity of biochemical reactions that keep a cell alive and functional; ordinary enzymes would be inhibited by such high salinities, but those of halobacteria prove actually to need that salt: they do not work, or work only poorly, unless they have salinity levels that high in the surrounding solution. In some way that is not wholly clear, potassium ions in particular interact with the cell's enzymes and enable them to function.

Potassium ions are not scarce in salt deposits, but they are vastly outnumbered by sodium ions. Halobacteria have developed methods of taking up potassium ions selectively and keeping unwanted sodium ions out. The potassium ion concentration within a halobacterium may be 1000-fold greater than that outside it. Such processes need a membrane and uptake machinery of considerable subtlety, to concentrate potassium and exclude much of the sodium. Halobacteria also need a sort of pumping system to expel any unwanted sodium that might get in. Pumps and selective uptake devices need fuel – biological energy – and halobacteria, unlike *Dunaliella*, are not able to photosynthesise energy-yielding organic matter from carbon dioxide. They need pre-formed organic matter as food, and supplies of this are likely to be intermittent. However, their membranes do possess a unique way of obtaining energy from light.

In their membranes are patches of purple pigments called rhodopsins. These are light-sensitive. Actually, they are chemically related to pigments in the retinas of our eyes, the light receptors which enable us to see. Not that halobacteria can see; in their case the pigments capture light energy and use it to generate a substance called ATP, which is the universal source of biological energy in cells. (I shall not bother with ATP's full name here.) In the process of generating ATP, ions are swapped between the interior and exterior of the cell such that most of the sodium ends up outside and potassium is retained: light helps the ion pump work.

This is a special kind of energy-generating photosynthesis, unaccompanied by carbon dioxide fixation, and halobacteria are the only organisms in which it is known.

I mentioned earlier the pinkness of many salt deposits, a colour partly due to halobacteria living in the saturated brine between the crystals, partly to entrapped corpses. Perhaps the most surprising discovery of recent years is the fact that halobacteria can actually colonise rock salt itself. When salt crystallises, it ought to form wholly solid, cube-shaped crystals. However, in practice particles of dust, sand, and even whole microbes, interfere with the crystallisation process in such a way that inclusions of saturated salt solution become trapped within the crystals. In the late 1980s, Dr. W. D. Grant and his colleagues at Leicester University showed that, if sodium chloride crystals were grown in a culture of halobacteria, not only were microbes entrapped in the fluid inclusions, but they remained alive there, inside the crystal, for several months. So they examined a natural subterranean salt deposit, one which dated from the Permian period, over 230

million years old, and there they found fluid inclusions within
the rock salt, too. Amazingly, a few of these (six out of some 350
samples) contained living halobacteria, which they could cultivate
back in the laboratory. Naturally, they sterilised the surfaces of
sample crystals, and handled the material with the utmost microbiological precautions, for they were well aware that a laboratory
working on halobacteria has a higher probability than most of
finding halobacteria in its samples, simply by accidental contamination! The precautions they took allayed such criticisms, so they
could ask the question, how long had those halobacteria been
there? The answer is still uncertain: it is barely conceivable –
note my cautious use of the term 'barely' – that they had been
alive, but dormant, ever since the Permian period. But they might
well be direct descendants of such microbes, clones which had
lived in their mineral prison for millennium upon millennium,
constantly recycling organic matter which had become entrapped
with them. Or they might have been sustained in exceptional
instances where minutely slow permeation of the crystals with
extraneous water brought in organic matter. In either case, they
displayed remarkable hardiness in what is a truly harsh and
restricted environment.

Halobacteria belong to a special class of bacteria known as
Archaebacteria, which are actually totally different organisms
from regular bacteria, even though they look much the same.
Many of the more exotic bacteria in this book prove to be archaebacteria, and I shall return to them often. As far as highly saline
environments are concerned, these salt-loving archaebacteria
have developed what is effectively a new physiology to deal with
their world: unlike all other organisms, they admit salt to their
interiors, but they do so on their own terms. They accept the
consequent dilution of their intracellular water and the slug-
gishness of water movement which must ensue (they do grow and
metabolise rather slowly compared with most other bacteria), and
they have special kinds of proteins and cell membranes adapted
to high salinities. Even their genetic apparatus is oddly unstable
– though I shall not enlarge upon that because there is no reason
to believe that it has anything to do with their special habitat.

Their reward for their physiological eccentricity is that they have a habitat almost exclusive to themselves, sharing it with just a few other microbes, principally *Dunaliella*. Indeed, *Dunaliella* must be a very useful colleague, since its photosynthesis, with concomitant starch or glycerol formation, is a primary source of organic matter in the food chains of their salty world. Halobacteria would be in poor shape without *Dunaliella*, just as we should be lost without plants. The price they pay is that they are salt addicts: they cannot do without salt and, like many an addict, they need more than would kill most other organisms. For them, the normal waters of this planet, marine or fresh, are lethal.

6

Corrosive and slippery places

Pffffff ...

My daughter Lucy inherited diverse qualities from her devoted parents. I like to think that they were mostly desirable, but I must acknowledge that genetics slipped up in one respect: she is by no means at ease with matters of chemistry. At one stage in her career she took an elementary course in agricultural science, a subject spanning biology, physics and chemistry; in the context of soil quality, she was obliged to learn about acid versus alkaline soils – very important matters to the farmer. Acidity and alkalinity are measured as a quantity called the pH, and to the uninitiated, pH is a strange, even alarming, symbol, partly because, with the capital letter at the end, it seems to be the wrong way round. Also it is pronounced 'pee-aitch', not 'pffff'.

One evening she consulted me, and our conversation went something like this:

'Dear and all-knowing father,' she said, 'I have a difficulty. What is pH?'

'Beloved daughter,' I replied. 'It is very simple. It is but the negative logarithm of the hydrogen ion concentration.'

My reply put an effective stop to any further attempt on Lucy's part to understand what pH is, though it solved her short-term problem most satisfactorily: she learned the phrase 'negative logarithm of the hydrogen ion concentration' by heart and reproduced it parrot-wise when required. She did well in her course.

Since then I have often wondered how I ought to have explained this quantity to a non-mathematical non-scientist. There may be some such among my readers, so let me try again, coming at it slowly, because pH not only matters to farmers, it is a fundamentally important quantity for all living things.

It all stems from a property of water. Water consists of molecules composed of one atom of oxygen attached to two of hydrogen: the familiar H_2O. The molecules which make up any kind of substance, solid, liquid or gas, are in continuous motion at ordinary temperatures; in solids they vibrate about a fixed point; in liquids they move about more freely. In a vessel of pure water, the molecules are forever bustling around, bumping one another in a random manner. Some may stick together briefly, only to be bumped apart again. The important property for present purposes, however, is that, at any instant, a few (one in every 555 million) have split. Each has divided, one fragment consisting of the oxygen atom plus one hydrogen atom, the other fragment being the remaining hydrogen atom: for these very few molecules, H_2O has become OH plus H. These fragments have strong, electrical charges; they are called ions, the 'hydroxyl' ion, as the OH is called, is negative, and the hydrogen ion is positive. The fact that their charges are opposite ensures that electrical attraction pulls them together and they recombine quickly, but no sooner do they do that than new ions are formed elsewhere in the liquid. That is why I said 'at any instant': as well as bumping around, sticking together and un-sticking, water molecules are also constantly 'ionizing' and re-associating.

Acids are substances which, when dissolved in water, react with it so as to increase the proportion of hydrogen ions. Vinegar, for example, is weakly acid because its main constituent, acetic acid, reacts with water so as to raise the proportion of hydrogen ions about 1000-fold. Formic acid, which is marketed commercially for de-scaling kettles, removing chalk encrustations from sinks and so on, pushes up the proportion 10,000-fold when one part is added to ten of water; and a really strong acid such as sulphuric acid, comparably diluted, raises the proportion about 300,000-fold. Water with all these hydrogen ions around is extremely reactive and must be handled with great care; a strong sulphuric acid solution is highly corrosive, able to dissolve metals such as iron and zinc, as well as to attack skin and give one a nasty acid burn. All strong acids behave in this way.

The three acids I mentioned, acetic, formic and sulphuric, produce their acidity in the same way: by generating hydrogen ions. This is true of all acids; in effect, the thing which makes acids be acid is the number of hydrogen ions they generate in water.

It follows that one can measure the acidity of a solution, whether in the laboratory, in rain, or on the surfaces of soil particles, as the concentration of hydrogen ions present. When they deal with concentrations of substances, scientists usually need to know what weights of substances are present and also the numbers of their molecules. For this purpose they use quantities called moles. This is not a misprint; in this sense of the word, a mole of a substance is the weight of its molecule compared to a hydrogen atom, but in grammes. A water molecule, for example, is 18 times heavier than a hydrogen atom, so a mole of water is 18 grammes. A mole of hydroxyl ions is 17 grammes, and of hydrogen ions is 1 gramme.

In a litre of pure water there is 1/10,000,000th of a mole of each kind of ion. 1/10,000,000 is an awkward number to do arithmetic with, even if it is written as $1/10^7$. (I remind you that 10^7 means 1 with 7 noughts after it.) In aqueous solutions, as I shall shortly make clear, the hydrogen and hydroxyl concentration may be far from equal; they may range from $1/10^1$th (which is the same as 1/10th) to $1/10^{14}$th of a mole per litre: an intolerable range for ordinary arithmetic. In 1909 a Danish scientist, Dr. S. P. L. Sørensen, suggested, and everyone agreed, that this sort of physical chemistry would be a lot easier if one expressed hydrogen ion concentrations using the superscript numbers instead of the actual numbers.

That is what pH values do. The superscript numbers are called 'exponents' by mathematicians, which is what the little 'p' of pH signifies, and the 'H' indicates hydrogen ions. A pH of 7 means $1/10^7$th of a mole of hydrogen ions are present in a litre of water; a pH of 3 means $1/10^3$rd of a mole per litre, and so on.

Mathematically, the pH scale is logarithmic: each whole number represents a ten-fold change, but because increasing numbers indicate decreasing hydrogen ion concentrations, it is called a negative scale. To those unfamiliar with logarithms, fractional values such as pH 6.5 are a mite complicated, because the in-between scale is also logarithmic – pH 6.5 is not half way between pH 6 and pH 7 but nearer 6 than 7 – but the important message is this: *low pH values mean many hydrogen ions and high acidity*.

Water, at pH 7, is taken as neutral, and any solution of lower pH is acid.

When acids generate hydrogen ions in water, they decrease the proportion of hydroxyl ions concomitantly, because they create extra hydrogen ions for the hydroxyls to combine with. In contrast to acids, there are substances which react with water in an opposite way: they generate hydroxyl ions. These substances are called alkalis. The hydroxyl ions they generate tend to combine with the hydrogen ions so that their concentration goes below $1/10^7$ of a mole per litre: a strong alkali such as caustic soda (sodium hydroxide) can generate a pH value as low as 14: lots of hydroxyl ions are then present but only $1/10^{14}$th of a mole of hydrogen ions for every litre.

Second message: *high pH values mean few hydrogen ions and high alkalinity*.

Water containing high concentrations of hydroxyl ions is also extremely reactive. Strong alkali breaks down fatty substances (soap is made by boiling fat with alkali), it dissolves protein, attacks wood, corrodes aluminium; even a weak alkali such as washing soda (sodium carbonate), which has a pH of about 10.5 as ordinarily used, feels slippery to the fingers because the hydroxyl ions attack the protective greases of one's skin, making soap of them. (Never use washing soda without protective gloves: de-greased skin is very susceptible to infection and other kinds of damage.)

pH and life

The Earth's surface is not neutral, it is mildly acid. By this I mean that the seas, rivers and lakes, like the water occluded in soil, are most commonly at around pH 5. There are many exceptions to this statement – otherwise I should not be writing this chapter – but pH 5 represents the most usual situation. It is slightly acid because some of the carbon dioxide in the atmosphere dissolves in ordinary water, and forms a very weak acid called carbonic acid. This is the substance responsible for shifting the pH of water from neutrality.

In contrast, the pH of the fluid inside the vast majority of living cells is scrupulously regulated close to pH 7.7: very slightly alkaline. The way in which it is regulated is a complex story which I shall set aside just now. However, the reason why it must be so regulated is easy to understand: all the enzymes within a cell, the catalysts which enable it to metabolise, grow and reproduce, perform their diverse functions by interacting with hydrogen ions, either direct or through a second substance. So their effectiveness depends very much on the concentration of hydrogen ions: on the internal pH. In a healthy, functioning cell, the amounts of enzymes, and their local pH values, are exquisitely adjusted, so that all enzymes function optimally. To this end, cells have elaborate mechanisms for excreting excess hydrogen ions and, when necessary, for generating or taking up hydrogen ions.

I wrote of a 'local' pH just now. Cells of plants and animals do not contain just an undifferentiated soup of enzymes; they have organelles within them, such as the nucleus. A pH of 7.7 is an average for the total cell fluid, and it may differ slightly as between various internal parts of a cell. This matter becomes important when one thinks about a cell as tiny as a bacterium, and for a remarkable reason: there is not room for many hydrogen ions anyway.

To explain this point I must do a sum, and to make it easy I shall round off the numbers involved. I pointed out that at pH 7 there are very few hydrogen ions relative to the number of water molecules; nevertheless there is still an astronomically large number in, for example, a litre of pure water. It is known to be about 6×10^{17}. But bacteria are exceedingly small; as an example, *Klebsiella aerogenes* is a common, rather ordinary, species of bacterium which lives in soil and pond water; it has an internal volume of about a cubic micrometre ($1 \, \mu^3$), which is $1/10^{15}$ of a litre. From which it follows that, if the cell fluid of *K. aerogenes* were at pH 7, there would be only about 600 (6×10^2) hydrogen ions in it. But in fact the pH is higher than that, 7.7, which brings the number down to about 200. As far as the chemistry of the cell is concerned, this is absurdly few: in most parts of the cell, most of the time, there must be none at all. What it means is that hydrogen ions are continuously appearing and disappearing, as part of the multifarious biochemical reactions which take place within the cell, such that, on an average, only some 200 are free enough to behave as they would in an ordinary solution.

Microbial acid heads

Klebsiella aerogenes, like all bacteria, has a membrane round its protoplasm which regulates what goes in and out. Very few substances can flow freely through the cell membrane, but among those that can are dissolved gases and water molecules. The situation of water is surprising. With an internal pH of 7.7 but living in water at pH 5, *K. aerogenes* has some 300-fold more hydrogen ions outside its membrane than within. It manages to keep them out while letting un-ionised water flow freely back and forth: a miracle of atomic-scale sieving, especially when you realise that a water molecule has, at least in theory, 18 times the bulk of a hydrogen ion. (It is the charge on the ion which makes the difference, but I must not digress into that now.) All bacteria have this kind of problem, and all can sieve out hydrogen ions from permeating water. In fact, bacteria turn this ability to advantage: by pumping hydrogen ions out of the cell and readmitting them, they generate differences in electrical charge across their membranes which they exploit to obtain energy.

Ordinary soil and water bacteria can live comfortably in watery environments between about pH 8 and pH 5: a 1000-fold range of hydrogen ion concentration. Some prefer the more acid side of the range, others the more alkaline side. Fungi and yeasts tend to like a slightly more acid range. But many environments in nature become much more acid than pH 5, especially where fermentations are going on. Fermentation is living without oxygen; only microbes can do it, and it will be one of the subjects of the next chapter. For now, I shall explain that many kinds of bacteria, and a few other microbes, can live out their lives without oxygen, and when they do so they obtain the energy they need by breaking large molecules of organic matter into smaller ones. This activity is called fermentation (many will have met the word in the context of wine- and beer-making, wherein yeast, starved of air, ferments sugar to form alcohol). The products of bacterial fermentations are very often acids: lactic acid appears in the souring of milk, malic acid in wine-making; acetic, butyric and formic acids appear in silage, compost heaps and so on. These generate local acidity, pH values in the region of 2.5, which the microbes involved tolerate, though acidity often brings their activities to a stop. However, whole communities of bacteria and associated microbes exist which not only flourish in such acid environments but actually prefer them – some microbiologists call them 'acidophiles', but acid-tolerant is probably the better word. Generally speaking, both ordinary and acid-tolerant microbes maintain their internal pH values in much the same range as *K. aerogenes*, excluding hydrogen ions with great efficiency.

But even more dramatic levels of acidity can occur in nature, generated by a group of bacteria called the thiobacilli. These are exceptional in a variety of ways – indeed, they include representatives of some of the more exotic extremes of life, and they appear several times elsewhere in this book. The interesting feature here is that their metabolism leads to the formation of sulphuric acid, one of the strongest acids known.

The element sulphur is widespread in nature. Natural deposits exist here and there – huge ones in Texas and Louisiana are exploited industrially. Sulphur appears around volcanic emissions, deposited from volcanic gas, and it turns up in geothermal springs, such as those in Iceland and New Zealand. Sulphur atoms form part of living material, especially proteins, and when cells die and decompose, sulphur appears as a by-product of various decomposition processes. It is also formed in the atmosphere as a result of combustion, and later deposited on soil and water. Combustion and microbial decomposition of proteins ensure that tiny amounts of sulphur are present almost everywhere; sulphur springs and deposits provide large-scale sources.

Thiobacilli look quite ordinary under a microscope, but they have learned to use elemental sulphur as a source of biological energy. Just as we burn up organic matter – from our food – using oxygen, to obtain energy, the thiobacilli burn sulphur with oxygen to do the same thing. We make carbon dioxide from the carbon of the organic matter; thiobacilli make sulphur dioxide, though it actually appears as sulphurous acid (analogous to the carbonic acid which carbon dioxide forms in water). Sulphurous acid does not accumulate, however, because the thiobacilli make further use of it, oxidising it to sulphuric acid. Most thiobacilli can also use other sulphur compounds in addition, including the gas hydrogen sulphide (a common, rather poisonous, product of biological activity) and substances called polythionates and thiosulphates, both of which appear in small amounts in nature. The widespread distribution of sulphur ensures that small populations of thiobacilli exist everywhere, surviving although generally outnumbered by other bacteria.

Heaps of mined sulphur and volcanic sulphur deposits (provided both are damp), and sulphur springs provide habitats where thiobacilli dominate, and the sulphuric acid they produce not only excludes most other living things but can cause substantial economic damage. The story of how a certain type of thiobacillus earned the name *Thiobacillus concretivorus* (literally 'concrete-eating sulphur rod') is instructive. In the mid 1940s Dr C. A. Parker, an Australian microbiologist, published research on a problem with municipal sewer pipes. These were made of concrete, but after only a couple of years' use their roofs would crumble and fall in. It transpired that the sewage which they were carrying was releasing hydrogen sulphide (as sewage always does; it is a perfectly normal minor product of sewage breakdown). Being a gas, it was diffusing to the roof of the pipe, and there thiobacilli were oxidising it, first to sulphur, then to sulphuric acid. Concrete is based on calcium carbonate; this dissolves readily in even quite dilute sulphuric acid. So the concrete pipe roofs disintegrated.

Dr Parker isolated various kinds of thiobacilli from the corroding pipes, using a laboratory medium containing a few mineral salts with sulphur floating on top. He was amazed to find that one strain of the bacteria, as it multiplied, brought the pH of his culture down to less than 1: about as strong as the 'dilute' sulphuric acid used in chemistry laboratories. It was so strong that it would dissolve fragments of zinc metal. Yet the bacteria remained alive in the acid fluid. Impressed by their tolerance of strong acid, and by the rate at which they could decompose concrete, Dr Parker gave his strain of bacteria its evocative name. Sadly, it later proved to be the same as a species that had been discovered many years earlier, called *Thiobacillus thio-oxidans*, so his splendid name had to disappear from the scientific literature.

I handled cultures of that thiobacillus myself when I was a young research scientist. In my first job, Miss Mollie Adams, a senior colleague who was an expert on the microbiology of sulphur compounds, played a microbiological joke on me. She gave me a sample of the thiobacillus culture so that I could make a microscope slide of the organisms in it. This was in the days, before modern phase-contrast microscopy, when the correct procedure for looking at bacteria, as taught to bacteriologists, was to dry a tiny drop of culture onto a glass slide, then warm it momentarily to stick the dead bacteria to the glass, then stain the (often invisible) deposit wih a dye, and look at it under a microscope. I duly did so. I observed nothing: rarely had a slide appeared so clean. I tried several times wth no greater success; then I called for help. An amused Mollie explained: as the specimen dried on the slide, the water in the droplet evaporated to leave concentrated sulphuric acid. This, momentarily heated, digested the cells of thiobacillus completely, leaving no visible trace. The secret of success was to neutralise the culture with some alkali first. She had not mentioned that – I think she felt that this brash young scientist, fresh from Oxford, needed a little lesson in the problems of exotic microbiology . . .

T. thio-oxidans is mighty tough, as they say in the West. Almost as tough is another species called *T. ferro-oxidans*. This species inhabits sites where an iron ore called iron pyrites* is plentiful, such as natural pyrites deposits; waste heaps outside copper, lead or uranium mines; and subterranean strata of pyrites in gold and coal mines.

Iron pyrites is a compound of iron and sulphur and, though it is a rock, these bacteria manage to attack it; provided it is wet, they make it dissolve and convert some of the sulphur to sulphuric acid. They actually use the dissolved iron ions as a source of energy, too (how is a matter for Chapter 8). During the process, some free sulphur is formed, which other thiobacilli can share, adding to the acidity. The consequence is that wet pyrites is converted to a highly acid mixture of iron sulphate and sulphuric acid, while the bacteria obtain energy wherewith they make carbon dioxide from the atmosphere into organic matter. This they use to grow and to multiply, but when they die and decay it becomes available to more ordinary, though acid-tolerant, microbes, including yeasts. On all of these, if things are not too acid, certain kinds of protozoa graze. A web of food chains can develop round wet pyrites, sustaining an exclusive community of acid-tolerant microbes, all supported primarily by *T. ferro-oxidans*.

Rain, and soil alkalinity, often dilute or alleviate the acidity of small-scale effluents from pyrites, but a large pyrites dump, or a pyritic stratum in a mine, can become dangerously acid, and cause serious problems to mining engineers by corroding pumping

* Pronounced *pyraitees* because of its Greek origin.

machinery and equipment; and the scorched land around water emerging from a rain-washed pyrites heap can be an environmental catastrophe. But on the credit side, the exotic chemical activities of thiobacilli have been exploited commercially to concentrate useful metals, such as copper and uranium, from low-grade pyritic ores.

Both *T. thio-oxidans* and *T. ferro-oxidans* require acid conditions to grow: if you want to culture them in the laboratory, you must add sulphuric acid to your culture medium, the solution in which they are to multiply, until it reaches about pH 2.5 before inviting them to grow in it. They are true acidophiles. A modest kick-back for the microbiologist is that, for many purposes, the medium need not be sterilised: almost nothing else will grow in it! Several other species of the genus *Thiobacillus* are known which do not need quite such acid conditions, nor do they generate, or tolerate, such exceptionally high acidities. Even so, they flourish in the range between pH 5 and pH 2, conditions much more acid than most microbes – and higher organisms – can stand.

How do our two microbial tough guys manage? They actually face two major physiological problems, one of which arises because they live in water, in which their 'food', sulphur or pyrites as the case may be, is insoluble. How do they get at it? I must set that matter aside, too, for Chapter 8 and consider here the equally perplexing matter of their corrosive habitat. Are the insides of the bacteria that acid?

It transpires that they are not. Their internal pH, the pH of the protoplasm within their cell membranes, does make a small concession to all that external acidity. Ordinary cells, and bacteria such as *K. aerogenes*, have, as I said, the slightly alkaline internal pH of 7.7; the thiobacilli settle for an about ten-fold increase: their internal pH lies between 6 and 7. Not much of a concession: it means that the thiobacillus cell must often handle a more than 10,000-fold imbalance in hydrogen ions between its interior and exterior, while yet admitting water molecules freely, and also regulating the export and generation of hydrogen ions consequent on its ordinary biochemical processes. Thiobacilli have learned to do this successfully, and thereby enabled themselves to colonise terrestrial environments which no other creatures can inhabit.

But it seems that, in some way that is not clear, they have been obliged to adjust their physiologies in such a way that they cannot manage without masses of external hydrogen ions. One cannot but wonder at, if not necessarily envy, their biochemical proclivities.

Slippery ones

Let me turn to the opposite of acidity. The most striking thing about alkaline environments is, as I mentioned earlier, that they feel slippery. I mentioned the reason, too: that the alkalinity – which is due to hydroxyl ions, I remind you – attacks the greases of one's skin and makes soap of them. The process is slow, but it takes only a tiny amount of soap to make one's fingers feel slippery. This is not a trivial point, because one's skin greases are composed of fat, and fat is also an essential component of cell membranes. Even fairly weak alkali, around pH 11, will destroy the membranes of animal tissue cells, or of *K. aerogenes*. More dramatically, if you boil a suspension of *K. aerogenes* with caustic soda at pH 13 – which is a fairly strong solution – the cells vanish. They are not only cooked: but their membranes disrupt, to become dissolved largely as soap, and the contents, mainly protein, are digested and dissolved, too. Only a tiny residue of alkali-insoluble material remains.

What happens to cells at boiling point also happens at ordinary temperatures and in weaker alkali, only it takes longer. The upshot is that even moderately alkaline environments are lethal to most living cells: they disrupt and dissolve them.

Alkaline environments are rare on this planet. A few result from industrial activity, such as the effluent water of a cement factory, and a few rather esoteric natural sites exist: birds' nests, for example, become alkaline because the excreta of fledgelings decomposes to release ammonia, which is alkaline in water (typically between pH 10 and 11). Natural alkaline waters occur in special geological conditions; water that has come through chalk strata tends to be alkaline, about pH 8, and in a few places calcium-rich waters have higher pH values. But all are readily neutralised by such factors as acid rain, or acids formed by microbes. Water that is more stably alkaline is found in so-called soda lakes, which are not common but which are widely distributed about the world, e.g. in Africa, Russia, Latin America, China and Australia. They are lakes which, because of the geology of their water sources, have become saturated with sodium carbonate, and they usually contain a lot of sodium chloride (common salt) as well. Their pH values range fron 10 to 11.5 and higher organisms can rarely grow there. But microbes exist which can, and such lakes may become positive hotbeds of biological activity.

Among the bacteria that live in soda lakes are specialised archaebacteria which need both the salt and the soda: they cannot grow or even survive without both. I wrote in Chapter 5 about 'halobacteria', bacteria which can grow only in strong sodium chloride solutions; the soda lake organisms are relatives of the halobacteria which, needing alkali too, are even more exigent in their environmental requirements. They are called *Natronobacterium* or *Natronococcus* according to their shape.

What do they live on in such an unpromising place? In fact, a wide variety of more orthodox bacteria are also present, most of which need organic nutrients too, but among them are cyanobacteria, organisms which, like plants, use sunlight to make carbon dioxide into organic matter, generating oxygen at the same time. Their photosynthesis provides a primary input of organic matter into the biological food chains in soda lakes.

Dr W. D. Grant of Leicester University studied a soda lake in the White Highlands of Kenya, in which the cyanobacteria are of a coiled filamentous type called *Spirulina*; they form mats in which they are so thickly clustered as to be visible. (Incidentally, *Spirulina* species are quite common in salty lakes and are highly nutritious: not only do flamingoes eat them, but local people harvest them, dry them and make them into an edible cake. One can sometimes buy dried *Spirulina* in health food shops.) Photosynthesis by *Spirulina* can lead to massive green blooms on the water surface. In due time layers of growth become overgrown, older material sinks and dies, and other bacteria, consuming and decomposing the dead matter, use up all the dissolved oxygen in the sub-surface water. Different types of bacteria then develop, capable of growing without air (called anaerobes), and some form hydrogen sulphide. This substance, poisonous to most air-breathing creatures, actually encourages growth of yet other anaerobes, including a bright red, S-shaped bacterium which rejoices in the almost unpronounceable name of *Ectothiorhodospira*. This one is very important in many soda lakes because it consumes the hydrogen sulphide, using it for a kind of photosynthesis (one which, unlike that in cyanobacteria, does not lead to oxygen being formed). Thus it adds to the primary organic input of the lake and enhances the general biological productivity. So much organic matter is generated by the colourful photosynthetic bacteria, and such is the rapidity of its consumption by natronobacteria and other non-photosynthetic bacteria, that the biological productivity of Dr Grant's lake – which means the turnover of organic matter through biological food chains – exceeds that of a very fertile agricultural soil. It is a turbulent microbial cosmos, from which most other living things are excluded, its exceptional biological production being driven by the tropical sunshine.

Other soda lakes may have a greater diversity of cyanobacteria, or may be more dependent on photosynthesising anaerobes such as *Ectothiorhodospira*, and these may sustain differing populations of both archaebacteria and orthodox bacteria. But high biological productivity by an exclusively bacterial population seems to be a widespread feature.

Bacteria which are obliged to live in alkaline environments are called alkalophiles. Many types of more ordinary bacteria have learned to tolerate modest levels of alkalinity, pH 9 or so, for example; one finds them in chalky soils and waters. They are not true alkalophiles because they are happiest at pH values near neutrality. Indeed, even some fungi, protozoa and water shrimps can tolerate the mild alkalinity, in the region of pH 8.5, of limy or chalky water. But true alkalophiles demand a pH substantially above 8 and do best at pH values over 10. Many thrive at pH 11.5 and the record is held by an alkalophilic cyanobacterium, a species of *Plectonema*, which is said to be able to grow at pH 13: equivalent to a solution of caustic soda capable of seriously damaging one's skin.

Both alkali-tolerant and alkalophilic types have to overcome physiological problems. They must have cell membranes which are resistant to attack by alkali. In archaebacteria such as the halobacteria, this is easy: the 'fatty' molecules in their membranes are not real fats at all, and they resist alkali naturally. But orthodox alkalophilic bacteria do make use of fat molecules in their membrane structure, and it seems that their membrane fats are simply more resistant to attack by alkali.

Alkalophilic bacteria also need hydrogen ions within themselves, to enable their biochemical machinery to work, and this means that they must keep their interiors more acid than their exteriors. This is the opposite of the problem faced by the acidophiles; alkalophiles have to exclude hydroxyl ions because these would otherwise neutralise their hydrogen ions – as well as doing a lot of other internal damage. Just as acidophilic bacteria allow a little slippage in their internal pH, letting it be somewhat more acid than normal, so alkalophiles permit themselves somewhat more alkaline internal pH values: slightly in excess of pH 9.

Do you recall that calculation I did earlier of the number of hydrogen ions which there would be in a *K. aerogenes* cell of internal pH 7? It came to 600. In a bacterium of 1 μm^3 (cubic micrometres) with an internal pH of 9 it would be 6! When one thinks about such tiny numbers of atomic particles, ordinary ideas of physical chemistry go somewhat askew: in fact, many more hydrogen ions must participate actively in the cell's biochemical processes; the implication of such a low absolute number is that almost none is in free solution. But the organism's biochemistry proves to be adjusted as if that high pH were a reality: many of the enzymes that can be isolated from alkalophiles are uncommonly resistant to destruction by alkali, and also work best in the presence of alkali. An important commercial use arises from this fact: ordinary soap is alkaline; enzymes from alkalophiles which destroy carbohydrates, proteins and fats are therefore used in biological washing powders because they tolerate the alkalinity of soap or soap-like detergents where enzymes from more ordinary sources would not. The detergent enzyme market is said to be worth some $200 million annually.

Envoi

This planet is mildly acid, and most living things have adjusted to a life near neutrality, just slightly more alkaline than pure water. People eat or drink substances which have pH values ranging from about 3 (e.g. rough cider, pickles) to about 8.5 (a dose of sodium bicarbonate), but within our cells we, like most living things, scrupulously maintain a pH close to 7.7. From *K. aerogenes* to *Homo sapiens*, life processes take place at this pH. Bacteria, as so often, extend this limit, to between pH 6 and 7 in the case of acidophiles, to over pH 9 in alkalophiles. That is a 1000-fold span in hydrogen ion concentration, and presumably it brackets the limits within which terrestrial biochemistry can work. But they are by no means the pH limits for life. Cells in our stomachs secrete hydrochloric acid (down to pH 1) which our stomach lining resists; there are seaweeds with bladders which contain quite strong sulphuric acid (they can give unwary bathers quite a burn); ants, some other insects, and stinging nettles, can generate formic acid, the strongest natural organic acid, in specialised organs, doing themselves no harm but burning their enemies. There is abundant evidence that terrestrial living things can adjust to, and even learn to exploit, seemingly lethal extremes of pH. Bacteria have been especially successful, developing networks of food chains which allow a variety of other bacterial species to share all but the most extreme environments.

On planets elsewhere in the universe – I take it for granted that there are a great many – life may have emerged and evolved in environments of pH values far removed from our terrestrial near-neutrality. There is free sulphuric acid in the uninhabitable near-furnace which is the Venusian atmosphere, for example. Once established on a world which is wet and cool enough, though, there is no reason why creatures analogous to our acidophiles or alkalophiles should not have evolved to a high degree of biological complexity. If space travellers ever manage to visit planets outside the solar system (which I sadly doubt), one hopes that acid- and alkali-resistant clothing and equipment will be among their supplies.

7

Life without oxygen

Breathing air

People and other animals, in common with green plants and most of the microbes, depend utterly on air for their existence. Without its oxygen they suffocate and die. Our familiar world of aerobes, which is the general name biologists have given to oxygen-dependent creatures, comprises a glorious variety of living things, large and small, mobile and rooted. But they all have this crucial biochemical feature in common: their life processes depend ultimately on the energy they obtain by burning – oxidising, in technical jargon – their foodstuffs with oxygen. In higher organisms, the food they oxidise is always organic matter, but a few bacteria can make use of inorganic materials such as sulphur or iron salts instead. Biologists call the oxidation of foodstuffs 'respiration'; it provides energy for everything that needs to go on in the living cell. In plants, respiration is overshadowed by photosynthesis, a process which leads to formation, not consumption, of oxygen, while the plant makes the organic matter of which it is composed from carbon dioxide. But plant cells have to respire, too. It is just that, in daylight, they make more oxygen than they consume.

I must pause here and clear up a possible misunderstanding. In everyday parlance, and in medicine and medical physiology, the word 'respiration' means 'breathing': inhaling and exhaling air, as in the phrase 'artificial respiration'. That is a perfectly correct use of the word. But early in the nineteenth century, among botanists in particular, the word 'respiration' acquired the more specialised sense that I have just defined. It is logical, because the objective of breathing is to facilitate respiration in that specialised sense, and lots of organisms do not breathe in the way we do. I shall use the word in the specialised sense for this chapter.

Respiration, then, amounts to the burning of organic food to obtain energy, though obviously it is not the vigorous burning that one has in a fire. Heat is indeed generated, but most of the energy is stored in a chemical, ATP, which I must now explain.

ATP (short for adenosine triphosphate; too much of a mouthful to keep saying) is a phosphorus compound, a substance which is present in all living cells. Biochemists isolated and purified it in the 1930s, and it can nowadays be bought, a white powder, from laboratory suppliers. Cells possess enzymes which make ATP, and also enzymes which decompose it. The important feature of ATP is that when it decomposes it releases energy; each molecule of ATP carries a little pulse of energy which cells can make use of. In the simplest case, the energy emerges as heat; that is how warm-blooded creatures such as ourselves keep warm. But the energy can be used for other purposes: for movement, as when a muscle contracts or a microbe swims; to produce light, as when fireflies glow; to pump substances in or out of cells. Above all, however, it is used to build up the complex molecules of proteins, fats, carbohydrates, nucleic acids and other substances of which cells are made. In effect, the whole object of consuming food, of breaking it down and oxidising it during respiration, is to use energy stored in the molecules of foodstuff to generate ATP, which it can then use. Some biochemistry teachers like to say that ATP is the 'energy currency' of living things: cells are dedicated to making it from food, and to expending it to sustain, and improve, their lives.

It may seem remarkable that, at the level of obtaining and using energy, living things are all the same, but it is so. Yet it does not follow that the processes are simple; things prove to be mighty complicated once one asks how precisely ATP works: for example, how does energy from oxidising a sugar becomes incorporated into ATP? Or how does the energy from ATP become the mechanical energy involved in lifting something up? I shall not pursue such questions here.

But I must mention two more details. Sugars are composed of carbon, hydrogen and oxygen atoms. When cells, such as those which make up ourselves, use sugar to make ATP, they do it in two stages. First, they cause the molecules of sugar to decompose into smaller fragments, including some carbon dioxide. This actually releases energy; not a lot, but enough to make a little ATP. At this stage no net oxidation takes place: oxygen is not involved. Then the smaller fragments are oxidised, entirely to carbon dioxide. Curiously enough, oxygen is still not involved: enzymes remove hydrogen atoms from the sugar fragments, effectively leaving carbon and oxygen atoms from the original sugar to form carbon dioxide, though the process is less direct than that. The important point for the moment is that oxygen gas comes in only at the final stage, reacting with the hydrogen atoms, again in a rather indirect way, to form water. Because the process is indirect, the energy can be tapped off gradually by the cell, and special compounds are involved, among which are red proteins called cytochromes, distantly related to blood pigments. At this stage a lot of ATP is made: some ten times as much as came from the first, oxygen-free breakdown of the sugar.

The essential messages, then, are (1) that aerobic respiration involves stages which do not involve oxygen, followed by a terminal stage which does, (2) that the oxygen-consuming stage

yields lots more energy than the anaerobic stage, (3) the upshot
is that oxygen becomes part of the surrounding water.

People are used to the idea that aerobic respiration is the oxidation of energy-rich foods by the oxygen of air, but they rarely
notice that it is equally the conversion of gaseous oxygen to water.
But that aspect of respiration becomes important when one comes
to consider a special, though large, group of creatures which
share this planet with us.

Living without air

I refer to the invisible world of anaerobes, organisms that do not
need oxygen to exist. Indeed, many do not tolerate oxygen. They
are almost exclusively microbes, and they inhabit places where
oxygen is absent, or is present at vanishingly low concentrations.
You might expect such niches to be rare, given that a fifth of our
atmosphere is oxygen, but they are not, because there are many
zones in the inhabitable reaches of this Earth where aerobes
consume oxygen faster than fresh oxygen can penetrate, and
thereby create the conditions demanded by anaerobes. Large-
scale examples include offshore sediments in the sea, mud in the
estuaries of rivers, decaying leaves in ponds and lakes; polluted
soil and water, parts of sewage treatment plants, compost and
dung heaps. I shall say more about such environments later; in
all of them it is mainly aerobic bacteria which use up the oxygen
and enable the anaerobes to grow. It is generally higher organisms,
or their activities, that provide these suitable environments in the
first place: decomposing end products, such as litter from plants,
corpses and excrements from animals. And animals also provide
haunts for anaerobes within themselves: the first stomachs, or
rumens, of ruminant mammals such as goats, cattle and sheep,
contain seething masses of anaerobic microbes living in symbiosis
with their host. Anaerobes are also present in more ordinary
intestinal tracts, in vertebrates (including ourselves) and insects;
they even inhabit the interstices of teeth. And they can infect,

and multiply in, wounded tissue to which swelling has restricted access of oxygen.

Anaerobes do not respire oxygen, so how do they get energy?

They have, in fact, six main ways of generating ATP. No one species can use all six: mostly they can use one, sometimes two, very rarely three of them.

The fermentors

Fermentation is a process that has been familiar to mankind since before recorded history: we exploit it to make beer, wine, cheese and a hundred other fermented foods, and more recently the name has been applied to the mass culturing of microbes whereby antibiotics and other pharmaceuticals are manufactured. The underlying process is the simplest, and probably the most primitive, way there is of making ATP anaerobically. In principle, it is just the first stage of aerobic respiration: the splitting of an energy-rich molecule such as sugar into two (or more) smaller molecular fragments, including carbon dioxide, and the grabbing of energy thereby released. Parts of the second stage may take place – hydrogen may be among the fragments released – but cytochromes and their ATP-generating activities play no part. The nature of the fragments is very important as far as other organisms are concerned, be they microbes or higher organisms, since they may include useful products such as alcohol, acetone or glycerol, as well as gaseous hydrogen – more about them later. But as far as the anaerobes themselves are concerned, fermentation is a relatively straightforward way of gathering biological energy. The fact that it seems to be a terribly wasteful way compared with aerobic respiration matters little, because fermentative anaerobes colonise places where organic matter is plentiful.

I mentioned a few examples of the sorts of environment in which organic matter is plentiful enough to support the profligate ways of fermentative anaerobes. Within such environments the activities of fermentors are often of profound economic or environmental importance.

Compost heaps, invaluable to gardeners, are man-made models of processes which are central to the economy of living things. Consider what happens in a compost heap. Organic detritus from plants is attacked by a variety of scavengers: aerobic bacteria, fungi, worms and molluscs such as slugs. In a very short time their respiration consumes all the oxygen within the heap. A little oxygen may diffuse in, especially a centimetre or two from the heap's surface, but throughout much of it there is virtually none. So worms and molluscs migrate to better aerated zones; aerobic microbes and fungi become dormant, and anaerobes flourish. They break down the components of the detritus, the cellulose, starch, proteins and most other substances, into fragments, sugars or comparable fermentable molecules, which they then decompose further. Some of the smaller fragments can diffuse to the outer reaches of the heap, to be used by aerobes and so to help keep oxygen out of the way; some are acids, such as acetic and butyric acids, which stop certain of the anaerobes from functioning but favour others: the microbial population actually changes as composting proceeds. In effect, the whole mish-mash breaks down to carbon dioxide, hydrogen and water, with some nitrogenous and sulphurous products (coming from nitrogen and sulphur compounds in plant matter), together with some products (notably methane) formed by non-fermenting anaerobes – more about methane later on. Ultimately a residue remains of stuff which is attacked by microbes only very slowly: the compost.

This, on a small, domestic scale, is what happens all over the place, though sometimes with modifications of detail. The floors of natural – and managed – forests and woodlands become extensive compost heaps in autumn. Similar processes flourish in sediments formed by leaves falling into ponds or lakes, in the sand of polluted estuaries, or onto marine beaches, in the mud of polluted rivers and canals. Even types of soil which are rich in organic matter can become anaerobic and sustain comparable microbial activities; and, of course, man-made situations such as sewage purification plants are hotbeds of anaerobic fermentation.

All over this planet, then, anaerobes find oxygen-free niches in which to flourish and, in doing so, they play a major part in bringing dead organic matter back into circulation, for other living things to make use of: the carbon dioxide they produce is used by plants for growth, and hence is consumed by animals and by man; and their nitrogenous and sulphurous products also re-enter biological circulation. Anaerobes have been vital to the continuity and development of living things for millions, indeed, billions, of years, so it comes as no surprise that some higher organisms have domesticated their own populations of such microbes.

The most impressive example is one I have already alluded to: the ruminant mammals. Consider sheep and cattle. They eat mainly grass, but in fact they cannot digest it. Instead they sustain a population of largely anaerobic bacteria in their rumens and these digest the grass, fermenting its molecules of starch, cellulose and protein to form small molecules which the animals can assimilate. The animals actually live on a diet of a substance called butyric acid, which is the major product of the rumen fermentation, together with some other acidic products and, of course, dead rumen microbes. (The microbes and the ruminants live together as a symbiosis; I expand upon such relationships in Chapter 15.) The rumen is a natural fermentation vessel, but it is a rather inefficient one because some hydrogen gas is produced and escapes – thus wasting energy – and also because it is an excellent habitat for non-fermenting anaerobes, which divert some of the fermentation fragments into methane, which the animal cannot use. Horses, like ruminants, are herbivores, but they have no rumen. In these animals, and other ungulates, the anaerobes live in their hind guts, where they perform much the same function: decomposing grass into molecular fragments, most of which the animals can assimilate. In fact, all animals, and people, need stable populations of microbes, mainly bacteria and including diverse anaerobes, in their intestinal tracts. Though these microbes are rarely quite so fundamental to their nutrition as in the herbivores, they could not manage without them.

I alluded earlier to that quite different way in which at least one higher organism exploits microbial fermentation: the art, for art it largely is, whereby people make good wine and good beer, exploiting the ability of yeast to grow anaerobically and split sugar to yield alcohol and carbon dioxide.

Yeasts, of which there are several kinds, do something of the sort in the wild. Yeasts are micro-fungi, organisms closer in character to plants than to bacteria (though, unlike plants, they require organic matter in order to grow and multiply). In nature, yeasts find sugar in such places as the nectar of flowers, the sap of some plants, or in the juices of ripe or decaying fruit. There they flourish and, in little microcosms of their own making, they use up the dissolved oxygen and then make their habitat alcoholic. The sleepy wasp of a late warm summer is not necessarily tired; it is probably drunk.

The lifestyle of yeasts illustrates a point about anaerobes which I have not mentioned before. Yeasts can grow perfectly well in air, respiring oxygen and generating energy efficiently. They have their quota of those cytochromes I mentioned earlier. But when the oxygen runs out, they can switch to an anaerobic metabolism, putting their cytochromes on hold and fermenting their food. Thus they have the best of both worlds, so to speak. Quite a number of bacteria can do this too. Microbes which can grow anaerobically or aerobically as circumstances dictate are called 'facultative anaerobes', and anaerobes which cannot cope with oxygen at all are called 'obligate anaerobes'.

Using sunlight

As far as having the best of both worlds is concerned, there is a group of bacteria which goes one better than yeasts and such facultative anaerobes. Plants, as I reminded you earlier, use sunlight to make the organic matter of which they are composed from carbon dioxide, so they do not need organic food. Naturally, they need the energy of ATP to build up that organic matter and sunlight provides energy to make that ATP. As part of the whole operation they generate oxygen gas from water. In effect, plant photosynthesis consumes carbon dioxide and makes oxygen, thus being the reverse of respiration, which consumes oxygen and releases carbon dioxide. The detailed mechanisms are very different indeed, however.

There are several types of bacteria, all coloured, which can use sunlight for photosynthesis. Some are aerobes and conduct plant-type photosynthesis, generating oxygen from water, but another group, more interesting for present purposes, can only photosynthesise anaerobically, and they do not split water to make oxygen. Instead, they split organic matter, as if fermenting, but they make more ATP than they would if they were simple fermentors. When they photosynthesise, they are obligate anaerobes, but they win out because they can also grow aerobically if oxygen and organic matter are available, though they cannot then photosynthesise. So they possess the efficient ATP-generating system of aerobes in addition to an especially efficient anaerobic process.

Even odder, to my mind, is a group of photosynthetic anaerobes which couple their photosynthesis, not to the splitting of organic matter, but to the conversion of an inorganic compound, hydrogen sulphide, to sulphur or sulphates. These do not have the wide options of the regular coloured bacteria, because the great majority cannot grow at all when air is present (they are true obligate anaerobes); but, in using sunlight and hydrogen sulphide to obtain biological energy, they can do something which no other living organisms can do. They will reappear shortly in the context of sulphate-reducing bacteria.

Breathing rocks

I have outlined two of the six main ways in which microbes solve the problem of living without oxygen: fermentation and anaerobic photosynthesis. The next three ways have it in common that they involve generating energy from certain mineral substances dissolved in water. The mineral substances are compounds which contain oxygen atoms: principally nitrates, carbonates and sulphates; and the process of removing oxygen atoms from these is known technically as 'reducing' these substances. The microbes I am concerned with possess enzyme systems which enable them to do this: to detach those oxygen atoms and reduce them to water, just as respiration makes water from gaseous oxygen. The microbes which can do these things are totally unrelated to each other, as I shall tell; they are a very diverse assembly of creatures.

Nitrate reducers

The bacteria known as nitrate reducers are found among several genera of regular bacteria. They can always respire in a normal aerobic way, using atmospheric oxygen, not nitrate, if there is some around: they are 'facultative' anaerobes.

Nitrates are compounds which are present in soil and water all over the world, but they are rarely there in high concentrations. They originate in three main ways: (1) they are formed by chemical reactions in the atmosphere, brought about mainly by lightning and fire, and are washed into soil and waters by rain; (2) they are the end products of the decay and putrefaction of the nitrogenous components of organic matter; finally, (3) some manufactured nitrogenous fertilisers consist of nitrates or, if they do not, they consist of urea or ammonium salts which soil bacteria turn into nitrates almost as soon as they reach the ground. Nitrates are much the most important sources of nitrogen for plant growth and, therefore, for our food supplies.

Nitrates are compounds with one nitrogen atom attached to three oxygen atoms. What the nitrate reducers do is to remove oxygen atoms and use them for respiration, in a process resembling the way in which aerobes use gaseous oxygen, though it involves a different set of cytochromes. Thus they augment the amount of ATP they get from their food; they do nearly as well for ATP from nitrate reduction as they do when air is present and they can respire conventionally.

As a result of nitrate reduction, the nitrogen atoms are left behind, and it is this that causes nitrate reducers to have important and curious effects on both our economy and our environment. A minor example first: several species of nitrate reducer are not good at picking oxygen atoms off nitrates, and leave two of the three behind. They thus convert nitrates into compounds called nitrites. These can react with proteins, making them resistant to attack by other microbes and turning them pink. But carnivores such as ourselves find such protein perfectly edible; nitrite simply acts as a preservative. It is this property that has been exploited for centuries in the 'curing' of meat: bacon and ham are made by treating pork with potassium nitrate, in conditions in which nitrate reducers grow and make potassium nitrite. These days food manufacturers often bypass the bacteria, and use nitrites bought from chemical suppliers as preservatives for meat products (as you will see if you read the small print on proprietary sausages, canned meats and the like).

A few nitrate reducers convert the nitrogen atoms left over from their anaerobic respiration to ammonia, which is utilised by other anaerobes, or even by plants such as rice, whose roots penetrate into waterlogged, anaerobic soil. But the nitrate reducers which profoundly affect our planet's economy are bacteria which convert them to nitrogen gas. The effect of this operation is to make the nitrogen inaccessible to plants, as well as to the vast majority of microbes – including, incidentally, those that do this thing – because plants, animals and the majority of microbes are completely unable to make use of nitrogen gas. Though it constitutes some 80% of the atmosphere, nitrogen gas is of no metabolic value to the vast majority of living things.

The kinds of nitrate reducer which make nitrates into nitrogen gas have been given the special name of 'denitrifying bacteria' because, by removing nitrates from soil, they decrease its fertility, and by their activities in sewage processing and in composting they spoil what would otherwise be good fertilisers. If it were not for the existence of a fairly small group of bacteria which can bring nitrogen back into circulation from the atmosphere – they are called nitrogen-fixing bacteria and I shall write more about them in another chapter – most of the nitrogen atoms available to living things on this planet, in soil- and water-borne compounds, would long ago have ended up in the atmosphere, Evolution would not have got very far and you and I would not be here to reflect upon these matters.

The denitrifying bacteria have a poor image, but they mean no harm and there is hope for them yet. One of the great problems of modern intensive agriculture is the need to add tonnes of fertiliser to agricultural soils, some 90% of which must be nitrogenous. There is no escape from doing this at present: a third of the world's 5.5 billion people would starve if industrial nitrogen fertiliser were suddenly abandoned. In fact, the world's food production is perfectly adequate even for that huge population, and the fact that some people starve while others get too much, apalling though it is, is a socio-political matter, not an agricultural problem. However, sloshing all this fertiliser about the globe's surface does generate a different difficulty: the plants catch only about half of it, and the rest, having becomes nitrates of one kind or another, is washed out of the soil and joins nitrates from other sources in the sea, and in lakes, rivers and subterranean aquifers, in due course turning up in our drinking water. Though there is little evidence that this does any harm at present, it could do in the future, and official water-regulating agencies, such as that of the European Community, have rightly prescribed nitrate limits for drinking water, which water companies must not exceed. This is an area of global pollution which did not exist until the human population explosion of the twentieth century; it is one in which denitrifying bacteria could be helpful, by returning the nitrogen of unwanted nitrate to the atmosphere, and ways of deliberately exploiting them for water treatment are being developed.

Will-o'-the-Wisp

Completely different from the nitrate reducers are the carbonate reducers. They, too, are all bacteria, and all obligate anaerobes. They are, indeed, very sensitive to oxygen, the least traces of which prevent them from growing. Even experienced microbiologists find them very difficult to handle, unless they are accompanied, in their cultures, by aerobic bacteria which mop up residual oxygen for them. Of course, in nature there are always plenty of aerobes around to do just that.

Carbonates are compounds in which a carbon atom is linked to three oxygen atoms. They are extremely abundant minerals – chalk, the rock beneath huge areas of Britain, is calcium carbonate, and carbonates of various kinds are present in dust, in sea and river water, and in soil. Carbon dioxide becomes a carbonate (carbonic acid), when it dissolves in water. The carbonate reducers are bacteria which take the oxygen atoms from carbonates and use them to generate ATP in a process analogous to respiration. However, though the detailed way in which they make their ATP is not well understood, enough is known to be sure that it is not much like the respiratory processes of aerobes and nitrate reducers. As with nitrate reducers, the remaining atom, carbon this time, has to do something. No half-way products corresponding to nitrites exist, nor is elemental carbon formed; the bacteria in fact attach some hydrogen atoms to the carbon atom. In the majority – the carbonate reducers which are ecologically important and scientifically curious – the left-over carbon atoms become methane. This is a gas, a compound of carbon and hydrogen, which is also called marsh gas or natural gas, and it generally seeps away into the atmosphere.

Methane-producing bacteria are responsible for a remarkable number of ecological and economic activities. I can only sketch a few here. Natural gas is now a major fuel of Western society, consumed on a scale comparable to oil and coal. Much of it arose from the anaerobic decomposition of plant material during the early history of life on Earth, when great quantities became entrapped in subterranean rock. The same process goes on today in the sediments of ponds, lakes and rivers. Poke a stick into such muds, especially in early winter when fallen leaves are decaying; the gas which bubbles up is methane, with some carbon dioxide. Occasionally it may ignite, especially over bogs rich in decaying material, and the dancing flame is known as Will-o'-the-Wisp. These methane-producing bacteria are among the inhabitants of the rumens of herbivores about which I wrote earlier, and the methane they produce is belched out by the animals. They act similarly in the hind guts of horses and other ungulates, the gas then coming out of the rear end, as the equestrian community well knows. But we, mankind, are in there with them: a major component of human flatulence is methane, produced by courtesy of our intestinal anaerobes. These microbes we contribute, via sewerage systems, to sewage treatment installations, where, with other anaerobes, they assist in breaking down organic wastes. Efficient sewage works catch the methane and use it as a fuel.

Bacterial halitosis

My last group in this clutch of anaerobes are the sulphate reducers. Like the nitrate reducers and the methane producers, they are all bacteria. They cannot use gaseous oxygen, and, though it does not kill them, it prevents them from growing.

Sulphates are compounds in which one sulphur atom is surrounded by four of oxygen. The sulphate-reducing bacteria remove these oxygen atoms and use them for a kind of respiration that is rather like aerobic respiration in that it involves cytochromes, special types exclusive to this group of microbes, and it leads to the generation of ATP. A peculiarity of the process, however, is that sulphate reducers have to 'prime their ignitions', so to speak. They have to use up ATP to get the reduction process started: one molecule for every sulphate molecule they reduce. They initiate their respiration by making a compound of ATP and sulphate, called APS (adenosine phosphosulphate, some will want to know), and this is what they really reduce. Once sulphate reduction has got going, they recoup more ATP than they put in.

Partly reduced sulphur compounds analogous to nitrites do exist, but the bacteria take all the oxygen atoms off sulphates and form only sulphides. Among such partly reduced sulphur compounds are sulphites, which have only three oxygen atoms round the sulphur instead of four, and thiosulphates, which have three oxygens plus one sulphur around a central sulphur atom. These and some others occur in nature, together with elemental sulphur, and sulphate-reducing bacteria can usually reduce them as well. Though I shall deal only with sulphate reducers here, I should mention that other types of anaerobic bacteria exist which cannot tackle sulphate but which specialise in reducing sulphur, thiosulphates or sulphites to sulphides.

Sulphides generate the product which is the most distinctive and important characteristic of sulphate-reducing bacteria: a frightful smell. The bacteria usually form calcium or sodium sulphides, because calcium or sodium sulphates are their commonest sulphate sources, and both decompose spontaneously in water to form hydrogen sulphide. This is a gas which not only has an evil smell of bad eggs but is also a powerful poison for most living things.

Sulphates, especially calcium sulphate, are ubiquitous. The rock gypsum is calcium sulphate, so is plaster of Paris. Sulphates are present in soil, water (especially sea water) and dust, and supplies are steadily deposited from sulphur oxides in the air (which for their part come from burning coal and oil, and are among the harmful components of acid rain). So whenever an environment becomes deficient in oxygen – such as when a pond becomes polluted with organic matter, or when the interior of a compost heap begins to rot vigorously – there is plenty of sulphate about and the sulphate reducers can start multiplying. In doing so they make hydrogen sulphide, and thereby have two profound effects on that environment.

Hydrogen sulphide reacts spontaneously, albeit slowly, with gaseous oxygen and removes it; the reaction yields water and sulphur. So hydrogen sulphide helps to keep oxygen away from the neighbourhood of the bacteria, which helps them to persist. Being poisonous, it also kills most other organisms and their dead remains provide the bacteria with more food. As a result, a well-established zone of sulphate reduction can be a self-perpetuating system. Provided supplies of sulphates and organic matter last, it will support the lives of several types of specialised anaerobes, bacteria which can tolerate, or make use of, hydrogen sulphide, and share the smelly, noxious microcosm. Among these are the photosynthetic anaerobes which use hydrogen sulphide that I mentioned earlier in this chapter; they help things along by generating organic matter from carbon dioxide using sunlight. And on the outskirts of the microcosm, where hydrogen sulphide meets air, there grow other specialised types of bacteria which oxidise the sulphide to sulphur or sulphate, consuming oxygen and so impeding its access to the zone.

When sulphate reducers grow on a large scale, they bring about environmental changes which have numerous ecological and economic consequences. Again I can only skim over them, but some are quite odd.

A very obvious effect is the nuisance associated with the pollution of water, and less frequently of soil. The grace and beauty of Venice and Bruges are marred, particularly in late summer, by the gentle effluvium of bad eggs – of hydrogen sulphide – rising from their canals; tourists are discouraged, metalwork tarnishes and paintwork turns black (that is probably why Venetian gondolas are painted black). In the 1950s sulphate reduction was so rampant in the Thames estuary that the Thames Conservancy had trouble with damage claims by the owners of ships berthed in London's docks. The Thames has been cleaned up tremendously since then.

Here is a tale of how illogically science may advance. In the mid 1950s I was a young researcher studying sulphate-reducing bacteria and my boss, K. R. Butlin (a distinguished microbiologist), was consulted about a polluted, flooded claypit in New Malden, Surrey. He took me along to look. The water was black and stinking of hydrogen sulphide, and it seemed that local residents were threatening the Council with legal action. The treatments we could suggest – filling or pumping out the pit, pouring in masses of acid – promised to be slow and expensive. Then one day, before our remedies could be put in hand, the Medical Officer phoned us: overnight the water had turned yellowish and stopped stinking – would we come and look? We came as quickly as we could and were amazed: it was quite true. What had happened, it transpired, was that someone had, stealthily, at night, and quite illegally, dumped a lorry-load of soil contaminated with waste from a chromium-plating works, half in, half out of the water. Could this have killed the bacteria responsible for the nuisance? Back at the laboratory we did some tests and discovered that chromates were indeed powerful, specific inhibitors of bacterial sulphate reduction, effective in concentrations of a few parts per million. Thus was the Malden Local Authority's problem solved by an unknown, and probably unknowing, benefactor, and an effective, if not widely applicable, way of controlling such pollution introduced.

Sulphate respiration has been held to be one of the commonest biological processes on earth, though one is rarely aware of it. Drastic natural manifestations of their activities may occur: there are periodic eruptions of hydrogen sulphide, arising from sulphate reduction in decaying seaweed on the sea bed, off the coast in Walvis Bay, South West Africa. On one occasion the town of Swapokmund was invaded, according to press reports, by 'clouds of sulphurous gas' which 'blackened silverware and the clock face'; dead fish were washed up on the beach in heaps, and 'sharks came to the surface gasping on the evening tide'!

A homely instance is the story of black sand at the seaside. Around estuaries and harbours the sand is often nice and brown until you start to dig in it, when it proves to be black and rather smelly underneath. However, determined builders of sand castles learn that, if they carry on, their grey-black constructs will gradually turn sand-coloured, over about half an hour, and will stay that way until the sea washes over them, when the sand goes black again. Why? The answer is that such sand has a fair content of organic matter – from detritus in the river or harbour – so, beneath its surface, out of the way of air, sulphate-reducing bacteria use this to make hydrogen sulphide. This reacts with iron compounds in the sand, to form iron sulphide, which is black. The iron sulphide is in the form of minute particles which, on exposure to air, react spontaneously with gaseous oxygen to form rust-coloured iron oxides. Thus the black sand turns sand-coloured in air; once the tide comes in, the process is reversed.

This, in miniature, is how major deposits of iron pyrites – an iron sulphide ore – originated in nature: hydrogen sulphide produced by bacteria precipitated iron sulphide in massive zones and this later became the special kind of sulphide called pyrites. Other metal sulphide ores may have arisen in similar ways over geological time, but that is not certain. However, it is quite clear that the vast majority of the world's deposits of native sulphur were formed as a result of bacterial sulphate reduction. During a warm, sunshiny part of the Permian or Jurassic periods, huge seas, which covered, for example, much of present-day Texas and Louisiana, dried up. Hydrogen sulphide, formed by sulphate reducers in the rotting marine vegetation, became oxidised by air, perhaps helped by other bacteria, perhaps spontaneously, to the element sulphur. Something of the sort still occurs today around sulphur springs, such as those in Yellowstone National Park, USA, or on the North Island of New Zealand. Thus these bacteria formed basic raw materials of twentieth-century industrial society, for sulphuric acid, which is needed by an enormous variety of industries, is most cheaply made from sulphur or iron pyrites.

Oil and sulphur deposits almost always go together, and the 'sour gas' of oilfields is hydrogen sulphide, apparently largely formed by sulphate reducers. These microbes inhabit the oil-bearing strata, deep underground, as well as being present in drilling muds, injection waters, the neighbouring soil and so on. They cause immense problems to the oil industry, including corrosion (of machinery, pipe lines and metal installations – including offshore platforms), blockage of oil seams, toxic gas in storage chambers and so on. Even after the oil is refined, such troubles are not over. Those large shore-side tanks used for storing petroleum delivered by tankers have sea water beneath the petroleum layer, and sulphate reduction in that water can lead to contamination of the petroleum with hydrogen sulphide, which air will oxidise to sulphur. For many purposes this does not matter, but the old piston-engined aircraft required very pure petrol or else their fuel pumps corroded. It is not widely known that, at a politically sensitive period in the summer of 1956, the sulphate-reducing bacteria managed to ground most of the UK's military aircraft in Malaysia. These bacteria cause the oil industry much expense in research and control measures. It is but a modest compensation to recognise that, millennia ago, they probably played important parts in the formation of the oil itself and in its coalescence into the vast subterranean fields we exploit today.

There are several odd-ball aspects of sulphate reducers. Their activities can 'sour' natural gas stored in su'bterranean aquifers or in gasometers; they can ruin paper pulp, darkening it and causing it to yield patchy, grey paper. They can interfere with tanning operations, get into patent medicines and, of course, spoil badly-preserved sealed or canned food – in that case their presence is immediately, if repellently, obvious. They can also foul up the machine tool industry by contaminating and degrading cutting oil emulsions. But one of their mischiefs leads me to a new slant on their anaerobic way of life.

I mentioned corrosion of machinery just now. This is an economic problem which extends far beyond the oil industry; the sulphate-reducing bacteria can corrode iron and steel; not just trivially, but so seriously as to cause the water and gas industries millions of pounds' worth of damage annually to water and gas mains. The way they do this depends on a quirk of their biochemistry: they can use hydrogen gas as a source of energy for sulphate reduction. They are so efficient at doing this that they can pick up and utilise the molecular film of hydrogen that normally coats wet iron and steel surfaces. It is that film which protects iron and steel from rapid corrosion; when it is removed, the metal surface becomes accessible to attack by water or sulphides. Ferrous water pipes buried in wet, anaerobic soil can corrode and perforate in but a few years.

Most sulphate reducers can make use of hydrogen gas to reduce sulphates, and so generate ATP. Why, you may ask, did they develop this ability in the first place? The answer lies in the activities of those fermentative anaerobes I wrote about earlier. Because they often produce hydrogen, quite a lot is formed in anaerobic environments. Microbes which can catch it and use it have a splendid source of energy from which to make ATP. Among such microbes are the methane-producing bacteria, but the sulphate reducers are better at it and usually grab most of what is available.

I must stop my account of these smelly little beasts. They hold a special fascination for me because, as I indicated, I studied sulphate reducers myself for many years, during which time they have, so to speak, grown from a couple of microbiological eccentrics to a positive menagerie of species, comprising a variety of physiologies, but all strict anaerobes. To my mind they present a bizarre yet self-contained way of oxygen-free life. As a group they overlap with other anaerobes in their proclivities. Some are thermophiles (see Chapter 1) and only grow in hot places; a few can reduce nitrates instead of sulphates if need be; some can grow fermentatively when necessary, splitting an organic compound instead of oxidizing it with sulphate; one species conducts the only known fermentation of an inorganic compound: though it can reduce sulphate with organic matter, it prefers to obtain energy by splitting a partly reduced sulphur compound called thiosulphate into sulphate and sulphide. And at least one type is unique: it can grow with hydrogen, carbon dioxide and sulphate only, using the hydrogen to reduce both carbon dioxide and sulphate, and *still* gain some ATP. As a group, the sulphate reducers are remarkably versatile.

Exploiting iron

The respiration of the last three groups certainly does not involve air, but oxygen comes into the matter because it is oxygen atoms, taken from nitrate, carbonate or sulphate as the case may be, which the bacteria utilise. My final group has no truck with oxygen at all.

Their way of life depends on a quirk in the chemistry of the element iron which I must explain briefly.

An uncomplicated chemical element, such as sodium, forms a single series of compounds, such as sodium chloride, sodium sulphate, sodium nitrate, sodium hydroxide and so on: it forms only one compound of each kind. Figuratively speaking, a sodium atom can hook itself on to another atom, so forming chemical compounds, but it has only one hook. The element iron behaves differently. Because of a subtlety of its sub-atomic structure, iron behaves chemically as if it has three hooks, and can use either two or three of them at a time. Thus where sodium can hook itself to just one chlorine atom, forming sodium chloride, iron has to hitch up either two or three chlorine atoms. In effect, it behaves as if it were two different elements: it is able to form two different kinds of chloride, two sulphates, two nitrates, two hydroxides and so on. Which compound it forms depends on how the chemist sets about making it. One set of compounds is called ferrous compounds; the other is called ferric, and analogous compounds often differ considerably in their chemical properties. For example, ferrous chloride is stable when dissolved in water, but ferric chloride decomposes slowly in solution.

Actually, the quirk, as I have called it, of forming more than one series of compounds is not uncommon: many elements other than iron also behave like this, and some form three, four or even more series.

Ferrous compounds can easily be converted to ferric compounds by appropriate chemical treatment, and the reverse change can be brought about readily, too. The fundamental process whereby a ferric iron atom in a compound becomes a ferrous iron atom is, in fact, exactly the same process as occurs when an oxygen atom is reduced to form water in respiration. I shall not elaborate here on what that process actually is; the important point for now is that conversion of a ferric compound to a ferrous compound is a sort of reduction, and the reverse – ferrous changing to ferric – is a sort of oxidation, even though no oxygen is involved.

Ferric compounds, usually a mixture of several, are abundant in soils and sand, in marine and freshwater sediments, and as the brown deposit in iron-rich streams. Bacteria prove to know their chemistry well: just as types exist which obtain energy by coupling the oxidation of organic matter to the reduction of nitrate, carbonate or sulphate, so there are types which obtain energy by coupling it to the reduction of ferric compounds to ferrous. Actually, numerous types of bacteria, and even some fungi, were suspected of being able to do this as long ago as the 1930s, because ferric compounds become transformed into ferrous compounds in bacterial cultures quite easily, but close examination of the examples showed that the transformation was just a fortuitous side reaction of microbial growth and respiration, of no metabolic importance. Only in the late 1980s were bacteria genuinely able to obtain energy from such reactions discovered, inhabiting aquatic sediments. They consume the sorts of compound that result from microbial putrefaction processes – such as acetic or lactic acids – while reducing ferric compounds to ferrous. Though only a few species have yet been isolated – most still lack proper names – they prove to include both facultative and obligate anaerobes. Moreover, there is now compelling evidence that they are very important in the recycling of organic matter in aquatic sediments: they are better at assimilating organic matter than either the methane producers or the sulphate reducers, so wherever there is lot of ferric iron about, as may often happen, the iron reducers out-compete the other types. They get first bite, so to speak, at the available food, and others flourish only when most of the ferric iron has become ferrous.

Envoi

Methane is present in ancient geological strata. There is good evidence that sulphate respiration was going on long before it became rampant in the Permian and Jurassic periods, when the world's sulphur deposits were being formed. For over 2 billion years of the history of life on this planet there was little or no oxygen around in the atmosphere, but living things were generating and regenerating organic matter which fermentative anaerobes could use. It is tempting to suppose that the fermentative anaerobes, the methane producers and the respirers of sulphate are today making use of lifestyles which emerged, in some form, very early in the history of terrestrial life; and that these were dominant for much of this period. Where and when the iron reducers might have fitted in is still debated, but there was certainly lots of combined iron about on the planet's surface, some of which would have become oxidised despite the scarcity of free oxygen, so they, too, may represent very ancient ways of life. Anaerobes ruled the world until evolution threw up plant-type photosynthesis, a process which gradually made oxygen a substantial component of the Earth's atmosphere, and so initiated the dominance of aerobes in which *Homo sapiens* now shares.

8
Living on minerals

Fire and brimstone

'Since we are all damn'd,' said Blackbeard the Pirate one day, 'come let us make a little Hell of our own, and try how long we can bear it.'

He went down into the hold of his ship, having previously closed all the hatches, and filled a number of pots and barrels with brimstone, sulphur, damp powder, oily rags and so on.

He then set the whole lot on fire, and the whole ship's crew stood the resulting stink and smoke, half mesmerised, for an amazingly long time. At last common sense returned, and they rushed up on deck, bursting the hatches, and let the filthy reek out. Blackbeard himself was the last to leave.

So the story goes. Brimstone and sulphur are the same thing (the narrator's chemistry was faulty), and stocks were carried on board ship to be mixed as needed with nitre and powdered charcoal (or soot) to make gunpowder. Blackbeard was a contemporary of Captains Kidd and Morgan, all notorious pirates, and their early eighteenth-century variety of international vandalism provoked responses among their fellow Englishmen not unlike the antics of today's lager louts and football hooligans: silent admiration among the socially disenchanted, ineffectual dismay among the rest. And, as the story makes so clear, the pirates were quite as credulous and stupid.

In choosing precious sulphur against which to measure their manliness, Bluebeard and his crew were drawing on an ancient outlook: that sulphur is a devilish substance. To the credulous this makes sense: even its vapour makes the eyes smart, and when it burns, it melts to to a viscous, bubbling mass, its weak, blue flame emitting a choking stink. It is discharged from the bowels of the earth, by volcanoes or evil-smelling hot springs, and where it emerges, animals and plants die, replaced by a glutinous, vari-egated slime. Surely sulphur is the effluent of Hell itself, a fore-taste and warning of eternal fire and brimstone, the punishment for sin which none can endure!

In the mid nineteenth century, in Europe and the USA, the general public used to follow developments in science and tech-nology with an avidity that is rare today. Reports and articles which were quite technical and detailed, dealing with current discoveries in basic science, in engineering and medicine, appeared regularly in pamphlets and newspapers; editions of scientific books sold out quickly, and public lectures and scientific demonstrations were a popular form of entertainment; scientific controversies, such as a conflict with creationists generated by Darwin's ideas on how evolution worked, were read and discussed widely. In science, as much as in philosophy, politics and religion, a state of intellectual ferment existed which penetrated to all but the most deprived levels of society. The gutter press (there was already one then) had not yet restricted Western society's curiosity to matters of sex, scandal and violence.

Mediaeval ideas about the world, about life and nature, were crumbling. So it is odd that, despite a proven, if modest, value in medicine and agriculture, sulphur was slow to lose its hellish image. It remained a sort of satanic metaphor, especially among evangelists. How astonished, then, must the public have been to learn of the researches of a Russian bacteriologist, Sergei Winogradsky, who, working in France in the mid 1880s, described organisms whose lives actually depended on sulphur.

That the slime of sulphur springs consisted of microbes had been known for several decades, and that fact in itself was surprising because the water of a sulphur spring is rich in hydrogen sulphide, a dissolved gas which is quite as poisonous, to most organisms, as hydrogen cyanide. (Hydrogen cyanide is prussic acid: the murder weapon beloved of early twentieth-century detective story writers.) Some of the organisms composing the slime were red, brown or green in colour; others were white or of a watery translucence. Under the microscope the majority were clearly bacteria, of all shapes and sizes: some static, as single organisms or clumps; some swimming about rapidly; some relatively large and filamentous. Many could be seen to contain granules within their cells and, in the 1870s, these were identified as particles of sulphur.

The 1880s and 1890s were a glorious period for microbiology. Spectacular advances seemed to be taking place week by week, as microbiologists built on the groundwork of the great Louis Pasteur. But most researchers were concerned with the roles of microbes in disease. Winogradsky was working away from the main stream of the science, being concerned instead with the microbes which inhabited the natural environment: those in soil and water. He discovered many new types of bacteria, some conducting hitherto unknown chemical processes, but perhaps his major contribution to microbiology was his realisation that some groups of bacteria, instead of utilising the carbon compounds which ordinary creatures need as sources of energy for growth and multiplication, could use various mineral substances. And sulphur was the first mineral he reported on.

The route by which he reached his conclusion illustrates, in a sardonic way, the roundabout manner in which science so often advances. Winogradsky studied a genus of bacteria called *Beggiatoa*. Beggiatoas are filamentous creatures, clusters of which grow as white flecks on the surface of polluted water, such as on a smelly pond full of decaying leaves in late autumn. Beggiatoas also form part of the slime associated with sulphur springs. Under the microscope *Beggiatoa* is very distinctive because it is huge, by the standards of most bacteria. (At one time it was thought to be a colourless alga.) Though its filaments are able to glide slowly over wet surfaces, they adhere to glass. Thus Winogradsky could watch a group of filaments on a microscope slide easily, and allow water to flow past them without its washing them away. So he did just that. He perfused them with clean water containing dissolved hydrogen sulphide, and saw that, provided the solution of sulphide was not too strong, sulphur granules appeared inside their cells. These could only have come from a reaction of the hydrogen sulphide with dissolved oxygen. Then he changed to water free of hydrogen sulphide, and the granules gradually shrank and disappeared. Throughout these processes, which took many days, the bacteria throve, and it came to him that the bacteria were using the hydrogen sulphide, and later the sulphur, as food. They were oxidising both with the oxygen of air, just as our own cells oxidise our carbonaceous food. Instead of producing carbon dioxide, as we do, they produced sulphuric acid – a nasty substance in large amounts, but they formed it slowly and it washed away without doing damage to the cells. The bacteria were using sulphur and sulphide as sources of energy.

However, food needs to be organic, to supply the carbon and other elements of which cells are made. Sulphur is sulphur and nothing else; where did *Beggiatoa* get its carbon from, in water which was almost carbon-free?

Winogradsky did a few experiments with another group of bacteria, also from hot springs, whose source of carbon compounds was then just being discovered. They were smaller and not filamentous. They usually contained sulphur granules but, unlike *Beggiatoa*, they were coloured red. Today they are known as the red or purple sulphur bacteria. Winogradsky was well aware that a German scientist, T. W. Engelmann, had recently obtained evidence that they were capable of photosynthesis: that, like green plants, they could use light to generate organic matter from carbon dioxide. In Winogradsky's hands, the red bacteria behaved in most ways like *Beggiatoa*: they made sulphur from hydrogen sulphide, and used up the sulphur when the sulphide was exhausted, producing sulphuric acid, but they only did this in the *absence* of air. On the other hand, *Beggiatoa* was not photosynthetic, and it died without air. Winogradsky spotted the link between *Beggiatoa*'s behaviour and that of Engelmann's bacteria: if the coloured bacteria were using the energy of light to make organic matter from carbon dioxide, surely *Beggiatoa* was using the energy from burning sulphur in air to process carbon dioxide in a similar way.

The hypothesis that minerals such as sulphur could serve as energy sources for converting carbon dioxide into cell material was unheard of but, like many a stroke of genius, the idea was elegantly simple. It has been abundantly confirmed since then, as I shall tell in the rest of this chapter. Winogradsky named bacteria which could use either light or a mineral for this purpose 'autotrophs' (meaning 'self-feeders'); autotrophs have since been subdivided into 'chemotrophs' ('chemical-feeders') which use minerals, and 'phototrophs' ('light-feeders') which use light.

Ordinary green plants are phototrophs though, unlike the phototrophic bacteria studied by Engelmann and Winogradsky, they form oxygen during photosynthesis, not sulphur or sulphuric acid. But plants pay a penalty for their autotrophic talent: they cannot make use of pre-formed organic matter at all. Nor, seemingly, could *Beggiatoa*, nor yet the red sulphur bacteria. Moreover, other chemotrophs which Winogradsky studied in his later researches could not use organic matter either. So he decided that a definitive feature of autotrophs was an inability to use pre-formed organic matter.

The concept of autotrophy revolutionised microbiology. So there is a touch of irony in the fact that, as far as *Beggiatoa* is concerned, Winogradsky was wrong. More recent research has shown that, though *Beggiatoa* may obtain energy from sulphur or sulphide, it is not an autotroph: it does not make organic matter from carbon dioxide. It needs organic matter to grow. It happens to be good at scavenging traces of organic matter from natural water and from the atmosphere. Even a sulphur spring has a tiny concentration of organic matter (it comes from dead bacteria, vegetation, dead insects and so on), and so did the 'clean' water of Winogradsky's experiments. But Winogradsky was not alone in making mistakes. Plants make oxygen when they photosynthesise, and Engelmann devoted a lot of effort to proving that his photosynthetic sulphur bacteria did so too. His 'proof' was flawed, and in 1907 another German scientist, H. Molisch, showed clearly that they did not. However, Molisch made his own mistake, too. He criticised both Winogradsky's and Engelmann's studies of sulphur consumption in coloured bacteria, maintaining that it could not be important because it did not always happen: coloured bacteria grew perfectly well without sulphide or sulphur. His mistake arose because he did most of his work with a group of red bacteria which we now know is quite different: they do not use sulphur compounds, and they do need organic matter.

Thus did the science advance. Muddle heaped on muddle not infrequently leads to truth, and in the 1930s another great microbiologist, C. B. van Niel (then working in Holland) resolved the matter of the coloured sulphur and non-sulphur bacteria, and their relations with light and oxygen. Today we know of many kinds of autotrophs, and also that, though some autotrophs cannot make use of pre-formed organic matter, others can (Winogradsky missed out there, too). Our understanding of autotrophy and the diverse ways in which it works has added much to our appreciation of the capabilities of terrestrial life forms.

Sulphur bugs

A true chemotroph able to use sulphide in air, in the way Winogradsky thought he had discovered, was first reported by yet another German scientist, A. Nathansohn, in 1902. His culture was soon lost, but the organism was re-isolated in 1904 by M. W. Beijerinck, a very distinguished microbiologist working at Delft in Holland. Beijerinck called it *Thiobacillus thioparus*, a name which will cause classicists to shudder because it is a hybrid of Greek and Latin. Loosely translated, it means 'sulphur-splitting sulphur rod', and indeed it was small, rod-shaped, and would grow in the absence of organic matter in a dilute sulphide solution, or alternatively with a compound of sulphur called thio-sulphate. In either case it needed air and formed sulphuric acid; it also grew with elemental sulphur, but very slowly. In 1914 the existence of a super-bug which would grow rapidly with sulphur was reported, but it was not until 1921 that S. A. Waksman (later the discoverer of the antibiotic streptomycin) and a colleague isolated it. They named it *Thiobacillus thio-oxidans*, and it is a truly remarkable beast: it makes so much sulphuric acid from the sulphur that nothing else can grow in its cultures, and the culture fluid can dissolve iron, zinc or concrete. It featured in Chapter 6 as the most acid-tolerant living thing known. Today many species, forming a rather diverse group called thiobacilli, are known, all able to grow chemotrophically in air using sulphur or sulphur compounds such as sulphide or thiosulphate. In addition, two highly specialised types of sulphur-oxidising bacteria have been isolated from volcanic hot springs; they are thermophiles (about which I wrote in Chapter 1), which require temperatures of 70 to 90° Celsius for growth.

One species of thiobacillus can grow without air provided a nitrate is present. Nitrates are compounds of oxygen and nitrogen, and this particular organism can break up nitrates, releasing nitrogen gas and using the oxygen atoms for sulphur oxidation. Most thiobacilli cannot use organic matter, but some retain the option: they can use either organic matter or sulphur, depending on what is available.

As I have said, thiobacilli are usually very tolerant of the sulphuric acid which they form. Other sulphur bacteria, not thiobacilli, are known which are less tolerant of acid but grow happily in places where another mineral, such as chalk, neutralises the acid as it is formed. The coloured photosynthetic sulphur bacteria, which have already featured in this chapter, are also not very acid-tolerant. Both groups prefer to use dissolved sulphides, such as sodium or calcium sulphides, rather than free sulphur as their energy source; such sulphides are plentiful in polluted waters, as well as in sulphur spring water. A remarkable type of thiobacillus, which also appeared in Chapter 6, can oxidise a totally insoluble mineral, a sulphide of the metal iron called pyrites. This organism attacks the mineral and makes use of both the sulphur and the iron as sources of energy for chemotrophic growth. Its name is *Thiobacillus ferro-oxidans*: 'iron-oxidising sulphur rod'.

More about how it uses iron compounds later. At the moment I want to discuss one of the most vexing questions raised by these thiobacillary proclivities.

Pyrites, as I just mentioned, is insoluble in water: its molecules do not enter solution at all. Sulphur, too, is not only insoluble in water, it is also water-repellent: it floats on water without getting wet. All bacteria, without exception, live in water and take their food from water; they have a cell wall and a cell membrane separating their protoplasm from the exterior, and they have nothing equivalent to a mouth or teeth: nutrients have to diffuse, or be transported, into their cells in solution. How, then, do thiobacilli even start to consume such inaccessible materials as sulphur or pyrites?

We still do not know for certain. However, we do know that, in both cases, the bacteria have to be in close contact with the solid surface and, in the case of sulphur, the bacteria somehow make its surface wettable. It seems likely that the sulphur and pyrites are attacked in different ways.

Consider sulphur. One explanation goes as follows. Sulphur has an appreciable vapour pressure – one can smell it – and though it does not dissolve in water, sulphur vapour diffuses through water just as it diffuses through air. So it enters the cell as a very dilute gas. Once inside, it can react with organic sulphides in the cell to form compounds called polysulphides; from these the organisms could form soluble sulphide, which they use easily. Alternatively they might somehow capture the sulphur vapour in an enzyme and attack it directly with oxygen – indeed, an enzyme capable of reacting sulphur directly with oxygen has been found in one of the volcanic hot spring types, but ordinary thiobacilli seem not to have it. However, the very fact that a cell can take in sulphur vapour and consume it would encourage more vapour to diffuse towards that cell: thiobacilli at the surface of a sulphur particle accelerate its normal evaporation. I like that explanation myself, but I am well aware that it is not conclusively established.

Now consider iron pyrites. This is a conventional rock and has no vapour pressure, so the sulphur explanation will not work. However, pyrites is attacked by sulphuric acid, to form a mixture of sulphur, iron sulphate and hydrogen sulphide. The sorts of thiobacilli which attack pyrites cannot grow unless the environment is more acid than most living things tolerate, and this provides the clue. It is highly likely that they cannot attack pyrites at all until it is already acid enough to begin to decompose anyway. Small amounts of sulphur are almost always present along with pyrites and, if the thiobacilli generate enough sulphuric acid from this to start decomposing the pyrites chemically, they can then accelerate the process by utilising the new sulphur to make yet more sulphuric acid. Thus they obtain one of their energy foods: sulphur. As I said earlier, they actually exploit the iron sulphate which is formed, too; obviously I must not postpone that story any longer.

Living on iron

Compounds of iron are very common on this planet's surface, in soil and sand, and dissolved in water. Streams in which stones, rocks, and even weeds, are coated with a brown deposit of iron rust are a familiar sight in localities where the water flows through or over iron-rich zones. Water coming out of peaty bogs can be especially rich in dissolved iron, and thick deposits, called 'bog iron', may form at the point of outflow. From the Iron Age into mediaeval times, bog iron was exploited as a very pure iron ore, but exploitable deposits are now exhausted throughout the world. In such deposits, under the microscope, tangles of filamentous bacteria can be seen, and the idea that microbes are concerned in their formation goes back to the 1830s. The history of those so-called iron bacteria is another story of confusion gradually resolving which I shall not rehearse here; the upshot is that, among the bacteria present there is at least one genus, called *Gallionella*, which is an autotroph. It obtains energy from transforming dissolved iron in water and uses that energy to make organic matter from carbon dioxide. Other bacteria which do the same thing are almost certainly present too, but the position is complicated because dissolved iron is quite reactive chemically and can form rusty deposits spontaneously around, sometimes inside, numerous kinds of bacteria, and also some fungi, doing them no obvious good, but no obvious harm either.

How does transforming iron compounds yield energy? It arises from a subtlety in the chemistry of the element iron which I alluded to near the end of Chapter 7. There I told how certain bacteria, in order to obtain energy from carbonaceous food for growth in the absence of air, transform iron compounds. In their case the energy came from carbon compounds, and the transformation of the iron compounds only served to release that energy – rather as the transformation of oxygen releases energy from carbon compounds in our own cells. In contrast to the anaerobes of the last chapter, *Gallionella* needs air to grow, so it uses iron compounds in a different way. But it exploits the same quirk in the chemistry of iron. To save looking back, I shall repeat the explanation briefly.

Sodium, which is an element with an uncomplicated chemistry, forms a series of compounds, such as sodium chloride, sodium sulphate, sodium nitrate, sodium hydroxide and so on. It forms only one compound of each kind; they are called sodium salts (sodium chloride is common salt). The element iron has a more complicated chemistry because it is able to form two sets of salts. One set is called ferrous salts; the other set is called ferric salts; which compound iron forms depends on how the chemist sets about making it. Analogous iron salts generally differ considerably in their chemical properties.

Ferrous compounds can easily be converted to ferric compounds by appropriate chemical treatment, and the reverse change can be brought about readily, too. The crux of the matter as far as iron bacteria are concerned is that the fundamental process, at the atomic level, whereby a ferrous iron atom in a compound becomes a ferric iron atom is, in fact, exactly the same process as occurs when a hydrogen atom is oxidised by oxygen to form water. Again I shall not elaborate on what that process actually is; the essential point for now is that changing ferrous iron atoms, in compounds, to ferric is a sort of oxidation, even though no oxygen is involved, and like many an oxidation, it yields energy.

The amount of energy released is not great compared with what our own cells gain when they oxidise glucose, for instance, so iron bacteria have to use a lot of ferrous iron to gain a reasonable amount of energy. This is why the rusty deposits in iron-rich waters are so bulky: it takes relatively few bacteria to build up quite a mass of rust. Iron compounds are very common in soil; when iron becomes dissolved in water it is usually in the form of ferrous compounds, most often the ferrous salts of organic acids, and bacteria such as *Gallionella* oxidise these to ferric salts. These are not stable in ordinary water, and they promptly decompose to give the rust: a mixture of ferric oxides and hydroxide. *Gallionella* uses the energy of the oxidation step to convert carbon dioxide to its own organic matter: it is a true chemotroph.

Gallionella is one of the few bacteria which have a somewhat plant-like form: it adheres to sufaces, grows a stalk and forms branches. The special shape is probably not a coincidence, because even in iron-rich water the actual concentration of dissolved iron is low. It must be advantageous to *Gallionella* to be well anchored in the same place and thus have its iron supply constantly renewed as water flows by.

Now I can return to the iron-oxidising sulphur rod, *Thiobacillus ferro-oxidans*, of which I wrote at the end of the last section. Once it has initiated the decomposition of pyrites by making its environment acid, sulphur and iron sulphate are formed together. The type of iron sulphate which is formed is ferrous sulphate. The thiobacillus oxidises this to ferric sulphate, which actually decomposes rapidly to rust and sulphuric acid, adding to the general acidity. Thus, as I wrote earlier, *T. ferro-oxidans* gets energy from oxidising both the sulphur and the iron component of pyrites, and it uses that energy to make its own organic matter autotrophically.

Pyrites is a kind of iron ore. Other metals, such as copper, tin, bismuth and uranium occur as sulphide ores and are decomposed by *T. ferro-oxidans*, though the metal component of the ore is of no benefit to the organism. Nevertheless, T. *ferro-oxidans* seems remarkably catholic in the minerals it will attack, and some strains are even more versatile, being able to use pre-formed organic matter, too, if it is available.

Nitrifiers

Thiobacillus ferro-oxidans, then, is an eater-up of minerals *par excellence*: it decomposes rocks, which even a mining engineer would class as minerals. But water-soluble substances such as chalk and common salt are also minerals and, because they dissolve in water, they are easier for microbes to tackle. In 1890, following up his ideas on autotrophy, Winogradsky described a new group of autotrophs which he had discovered: one which obtains energy from ammonium salts.

This needs explaining, because ammonia is actually a gas – many people have used household ammonia, which is a solution of the gas in water; they will be well acquainted with its pungent smell. Ammonia the gas is nitrogen hydride, a simple compound of nitrogen with hydrogen, and it is rather peculiar in that it behaves chemically as if it were a metal. I mentioned a few paragraphs ago that sodium reacts chemically with appropriate reagents to form salts such as sodium sulphate, chloride, nitrate, carbonate and so on. Ammonia does the same. Chemists call them ammonium salts. But the ammonium component of, say, ammonium sulphate can be attacked by Winogradsky's bacteria, provided air is present. They oxidise its hydrogen atoms to water, replacing them by oxygen atoms, so that the ammonium group turns into water plus dissolved nitrogen oxides, the latter in forms known to chemists as nitrites and nitrates. Bacterial oxidation of ammonia to nitrates is known to scientists as 'nitrification'; that bacteria could perform this reaction was suspected before Winogradsky proved it definitively. Two major genera of bacteria are involved: *Nitrosomonas* makes nitrites and *Nitrobacter* oxidises those to nitrates.

Oxidation of ammonia by oxygen yields energy, and the nitrifying bacteria, as they are collectively called, use that energy to make organic matter from carbon dioxide. They are true auto-trophs; they cannot make use of pre-formed organic matter. Indeed, they are so strictly autotrophic that they do not like organic matter: if there is a lot of it about, it seems to prevent them from growing and multiplying. At best, they multiply only slowly compared with most bacteria. Slow growth and sensitivity to organic matter were obstacles that Winogradsky had to over-come when he discovered and isolated these bacteria, and he thought intolerance of organic matter was a general feature of autotrophs because some of his thiobacilli seemed to dislike organic matter, too. But it is not so. The nitrifiers are indeed sensitive to pre-formed organic matter, but most other chemo-trophs and phototrophs are not and, as I indicated earlier, some types can actually make use of it.

You may well ask, where do nitrifiers get the ammonia they need to grow? The answer is, everywhere. When biological material decomposes and putrefies (i.e. when bacteria and other microbes break it down), the nitrogenous compounds – proteins, DNA and so on – ultimately release their nitrogen atoms as ammonia. This dissolves readily in water, but it sticks to ordinary soil, becoming absorbed on soil particles. Here the nitrifiers use it, converting it to nitrates. So the main habitat of nitrifiers is soil, and there a steady supply of ammonia is assured.

Nitrates are the principal nutrients that plants take up from soil: for most plants, nitrates are their sole source of nitrogen for growth. But as most gardeners know, the commonest artificial 'N' fertiliser is ammonium sulphate. It works because, when it reaches soil, nitrifying bacteria convert its ammonium component to nitrates.

The process of nitrification is not only important to gardeners and farmers; it is also a fundamental step in the cycling of nitrog-enous material among living things: it is the means whereby the nitrogen of biological excreta and dead matter is returned to soil for plants to utilise. Sometimes the bacteria perform too quickly for the convenience of farmers, and the nitrates escape into lakes and rivers before plants can catch them – but that is another story which I must leave for the present.

On a global scale, nitrifiers turn over some three billion tonnes of 'N' in a year. This is a huge amount – for comparison, plants turn over about thirty billion tonnes of 'C' as carbon dioxide. The nitrogen economy of the whole living world depends heavily on these organisms; it is curious that they are limited to only a handful of bacterial genera, all eschewing, detesting even, the organic matter which they do so much to generate.

Consuming hydrogen

Hydrogen is a gas, the lightest gas known. I am not sure that mineralogists would regard it as a mineral, but for this chapter I shall do so, because it is certainly not carbonaceous, even if it occurs in virtually all organic matter. One of the unsolved curiosities of the bacterial world is that a very large number of bacterial species possess an enzyme, called hydrogenase and not found in higher organisms, which enables them to consume hydrogen, oxidising it to water. Such bacteria do not necessarily need gaseous oxygen in order to do this: types exist which can remove the oxygen atoms of nitrates, carbonates or sulphates, and oxidise hydrogen with them; a few do not make water at all, but instead stick the hydrogen atoms on to an organic compound. The paradox is that, with most such bacteria, their ability to oxidise hydrogen serves no useful purpose – or no obvious one. Although energy is produced – as when hydrogen burns – most organisms make no use of it.

I shall not digress into the microbiologists' debate about what hydrogenase is for; the message for this chapter is that some thirty species of bacteria are known which have learned to exploit that energy. They are called 'hydrogen bacteria', and they live in soil and water. They are a diverse collection of organisms: the ability to exploit the hydrogen oxidation reaction seems to be scattered almost randomly among unrelated genera. A much studied example is a rod-shaped organism which swims about, called *Hydrogenomonas*; this can multiply autotrophically under a mixture of hydrogen, air and carbon dioxide, needing only a few essential minerals from its watery surroundings, exploiting the energy of the hydrogen–oxygen reaction to convert carbon dioxide to organic matter. It is just as much a chemotroph as the bacteria which do a similar thing with sulphur, ferrous iron or ammonium compounds.

Hydrogenomonas must have air in order to grow. I mentioned bacteria which can use nitrates, carbonates or sulphates in place of oxygen just now; the latter two types are, in fact, strict anaerobes: unable to grow if oxygen is present (they appeared in Chapter 7). All three types, in common with the great majority of other hydrogen bacteria, can use organic matter instead of hydrogen plus carbon dioxide, if it is available.

Hydrogen is a light, mobile gas which is barely soluble in water and soon bubbles out of solution, escaping to the atmosphere. So the question again arises, where do the bacteria get the hydrogen in the first place? The answer is that the decomposition of organic detritus, in locations such as soil, compost heaps and aquatic sediments, yields carbon dioxide and other products together with quite a lot of hydrogen. But one is rarely aware of the hydrogen because the bacteria catch it as soon as it appears. It transpires that the enzyme hydrogenase has a remarkable affinity for hydrogen, enabling the bacteria to trap it before enough appears to form a bubble. Indeed, in actively fermenting sediments, for example, the sulphate reducers and the carbonate reducers compete for hydrogen, the sulphate reducers being generally better at the game and becoming dominant as the supply of hydrogen begins to run out.

Hydrogen, then, is abundant in a variety of microbial microcosms and can support several modes of chemotrophic growth. This makes it the more odd that many hydrogenase-containing bacteria seem not to benefit at all from their peculiar enzyme.

Mixed-upotrophs

As humans, we are familiar with a major group of autotrophs, the green plants, and we take for granted their remarkable ability to use sunlight to create their own substance out of the air. Yet to my mind it is the chemotrophs, with their ability to exploit such bizarre materials as iron, sulphur or hydrogen, in some instances without recourse to light or oxygen, which illustrate most impressively the chemical versatility of terrestrial life. Indeed, one can divide the living world into two great nutritional groups. One is the autotrophs, which live by using either light energy (plants and coloured bacteria) or chemical energy (chemotrophic bacteria) to convert carbon dioxide into organic matter; the other is the rest, which consume the organic matter which the autotrophs make, either direct, as do herbivores, or at one or two removes, as do carnivores as well as many microbes. That some creatures belong to both groups – carnivorous plants, for example, or hydrogen bacteria – does not invalidate the division. However, a different kind of cross-over, so to speak, exists, which is found only among bacteria.

Green plants, and most photosynthetic bacteria, make organic matter, with the aid of light, from carbon dioxide, and only from carbon dioxide. I mentioned early in this chapter coloured sulphur bacteria which, while oxidising dissolved sulphides to sulphur and sulphuric acid, also photosynthesise their own organic matter from carbon dioxide. A few decades ago, biochemists discovered that certain types, called green sulphur bacteria, could use a simple organic compound, acetic acid, as well as carbon dioxide, and photosynthesise this into complex organic molecules. Within a few years a wide variety of bacteria, phototrophs and chemotrophs, proved to be able to do something similar: they could use the energy of light, or of a chemical reaction such as sulphur or hydrogen oxidation, to assimilate acetic acid or other simple organic compounds into themselves. However, they had to assimilate an equal amount of carbon dioxide at the same time.

This ability is a sort of half-way house between autotrophy and our kind of nutrition; it has been called mixotrophy. Some microbiologists have speculated that mixotrophy represents a very primitive kind of autotrophy, one stemming back to the earliest period of life on this planet, when organic matter is thought to have been abundant and when it would have been advantageous for an organism to assimilate efficiently what was available. If so, why can plants not do the same today? It would surely benefit them. For myself, I am inclined to think that mixotrophy is a late adaptation of autotrophs to life amongst the abundance of organic matter which is available on the planet's surface today. For as biological matter putrefies and decomposes, small organic molecules such as acetic and other acids always appear, alongside carbon dioxide and hydrogen. Many bacteria grab these small molecules and oxidise them further to make yet more carbon dioxide, but organisms which can make use of them direct, without going through carbon dioxide, are, I think, at a competitive advantage.

Be that as it may, the existence of a third kind of 'trophy' opens new vistas for the evolution of terrestrial-type life. On this planet the carbon cycle, whereby organic matter becomes carbon dioxide which then becomes organic matter again, seems absolutely fundamental to the existence of living things. Yet in principle carbon dioxide could be much less important to life's persistence, and elsewhere in the universe evolution may have generated communities in which the formation of free carbon dioxide has been dispensed with altogether.

9

Exotic menus

Poisonous plants are hazards that the country dweller learns about early in childhood. And rightly, because the plant kingdom produces a tremendous variety of poisons, many of which have been exploited from earliest times for medical, veterinary and warlike purposes. Such widespead toxicity is logical, of course; poisons are most effective deterrents of herbivores and confer evolutionary advantage on the plants that make them. Among the army of such noxious plants is a South African shrub called gifblaar,* which kills cattle if they happen to browse on it. It is of special interest to biochemists because the poison it produces, called fluoracetic acid, has a molecule very like that of an ordinary, harmless organic substance, acetic acid (the sour component of vinegar). Fluoracetic acid has since been found in about 40 plant species, mostly from the southern hemisphere.

---

* *Dichapetalum cymosum.*

Chemically, the fluoracetic acid molecule is very simple. Like acetic acid it is a compound of carbon, hydrogen and oxygen, but one hydrogen atom has been replaced by an atom of the element fluorine. Most compounds containing fluorine, organic or inorganic, are poisonous, so fluorine compounds are rarely found in living things – the principal exception is calcium fluoride, an inorganic compound of calcium and fluorine (as its name suggests), which is a minor but important component of healthy teeth.

Acetic acid is formed in small amounts from most kinds of food during their digestion, and it is consumed readily by the cells of our bodies. Once acetic acid gets into the bloodstream, or is generated in a cell, it rapidly ceases to be the free acid and becomes a salt, such as sodium acetate. So I shall refer to acetic acid in metabolism as 'acetate' henceforth. Similarly, fluoracetic acid, once it gets past the stomach, becomes fluoracetate.

During the 1940s, Sir Rudolph Peters, a leading biochemist of the day, and his colleagues discovered the reason why fluoracetate is so very poisonous: cells actually mistake its molecules for those of acetate. When fluoracetate enters a cell, it is put through the first few of the reactions which acetate ordinarily undergoes, which would cause it to be converted into citrate. Fluoracetate gets converted into a substance called fluorocitrate. The enzyme which would normally metabolise citrate, however, makes no mistake; it cannot handle fluorocitrate. But the fluorocitrate molecules stick to the enzyme, and inactivate it. The citrate enzyme is crucial to respiration, so the cell, figuratively speaking, chokes and dies.

Fluoracetate kills all kinds of air-breathing organisms, from microbes to people (except the plants that make it; one wonders how they manage), and there is no antidote. It is an easy material to make in a laboratory, and it has proved useful commercially as a reasonably cheap pesticide, though it is not often used today because of the risk of killing unintended victims. (Professionals still use it sometimes for killing insects or rodents.) For use as a pesticide it is marketed as a derivative called fluoracetamide, which breaks down into fluoracetate when the victim eats it.

In 1963 considerable public anxiety was generated when some

fluoracetamide was released accidentally from a factory in the South of England, contaminating the grass in neighbouring fields, actually killing local cattle and putting people at risk. What could be done about it? After much consultation, the authorities decided that the only reliable remedy was to remove the topsoil with mechanical diggers and to cart it away to a safe place, which was in due course done.

The saga went on for some weeks, and newspapers made much of the story. I remember discussing it with a young colleague, Dr. Michael Kelly, in our research laboratory, then located in Camden Town, London, and we thought what an effort digging and carting all that soil must be. How much easier things would be, we considered, if some kind of bacterium had been available which could decompose the fluoracetamide; it could have been sprayed over the buildings and soil to clean up the place. As far as we knew, none existed, so Kelly decided he would like to find one. It was not part of our research programme, but our boss worked some distance away in the Mile End Road and Kelly thought, and I agreed, that we could dovetail it in with our regular research. So on the way to the laboratory one day he collected some samples of mud and murky water from the River Thames and, at the lab., he added a little fluoracetamide and minerals, and left them in the warm. After about a week, to our pleasure, some bacteria grew in a few of the samples. So far, so good, but impurities in the water rather than the fluoracetamide might have been supporting multiplication of bacteria resistant to fluoracetamide's toxicity; Kelly's next step was to see if he could isolate a culture which would grow with just pure fluoracetamide and minerals. He succeeded. But impurities are difficult to exclude completely even in a laboratory (as I tell in Chapter 10), and to be doubly sure, he had to confirm that the fluoracetamide was being used up. It was. After a couple of months we knew that he had successfully isolated a culture of fluoracetamide-decomposing bacteria. By that time, of course, the accident that provoked his experiments had been handled the hard way: the topsoil had been removed and the land re-sown with grass. Well, we had known before Kelly started that success, if it came, would be too late for that particular incident, but about a year later a little paper came out in the science magazine *Nature*, and a sample of the pure culture (it was a variety of a common soil organism called *Pseudomonas*) was deposited in the National Collection of Industrial Bacteria for safe keeping. It is still there, but happily there has been no call to test it in practice.

The moral of that tale is that somewhere in the bacterial world there is something which will eat – I write metaphorically, for bacteria have no mouths – almost anything you can think of and, given patience, it is usually quite easy to find.

Plants and photosynthesising microbes do not eat, excepting the odd insectivorous plant and a few specialised microbes; instead, aided by sunlight, they make their organic matter from carbon dioxide and use it for energy, growth, etc. In Chapter 8 I wrote about some other exceptions: carbon dioxide-using bacteria which obtain energy by transforming minerals such as sulphur, ammonia, hydrogen and the salts of iron. These materials contain no carbon: they are solely sources of energy used by the bacteria, enabling them, like plants, to make their organic matter from carbon dioxide. But the food that animals and most microbes consume has two functions. One is to provide cells with energy, and the second is to provide the organic matter – the compounds of carbon – with which cells grow, multiply and maintain themselves.

The great majority of bacteria use pre-formed organic matter as food, just as animals do, and on the whole they tend to like the same classes of foodstuff: proteins, carbohydrates and fats. Even Kelly's fluoracetamide bug preferred more normal food. When consuming fluoracetamide, it detached and rejected the fluorine atom (which emerged as a fluoride) and used the rest of the molecule.

There are a few natural organic products which neither animals nor the majority of bacteria can consume. Examples are cellulose and lignin, substances which are to plants what bones are to animals: they provide a rigid – or fairly rigid – structure which gives the organism its form. Cellulose is the fibrous stuff of, for instance, straw, and lignin is a denser material that makes wood hard. Both are very stable substances at ordinary temperatures, and the few types of microbe that can consume and thus recycle them are extremely important in sustaining the turnover of organic matter for the rest of the living world.

Cellulose and lignin both consist of very large, long molecules containing carbon, hydrogen and oxygen. Neither is soluble in water. This presents microbes, especially bacteria, with a problem, because they have no way of taking up insoluble matter; they rely on molecules of their food being dissolved in water and soaking into the cell, usually helped by special proteins at the cell surface which absorb the molecules and help to transport them inwards. Bacterial cells have, on a molecular scale, thick overcoats: a cell wall and often an outer capsule. They cannot get their inner surfaces close enough to big, insoluble molecules such as cellulose to attack them. What they do is get as close to the cellulose as they can, often adhering to it, and then they excrete an enzyme called cellulase, a protein which partly digests the cellulose. This breaks it down into molecules which are soluble and which the cell can take up and consume. Of course, some of the molecules so released escape; other bacteria in the neighbourhood welcome them. A cellulose molecule is actually a long chain of sugar molecules linked together chemically, so cellulase breaks it down into quite a rewarding food for most bacteria.

Only bacteria and a few fungi make cellulase. Higher organisms do not. This is curious; it means that vegetarian animals, such as herbivores, snails and leaf-eating insects, cannot digest cellulose themselves. They solve their own problem by keeping a domesticated menagerie of bacteria in their intestinal tracts to break cellulose down, and they share the sugary product with the bacteria (they also digest some of the bacteria; I write in more detail about such associations between microbes and animals in Chapter 15).

The lignin of wood is a tougher molecule. It is a chain of substances called phenols, many of which are poisonous to bacteria as uncombined molecules. Indeed, many commercial disinfectants are phenols of one kind or another, distantly linked to those in wood because they come from coal tar, and coal was woody material several millions of years ago. Phenols are harmful to animals, though more so to bacteria, but on the whole plants do not mind them; they actually produce their own phenolic molecules as a protection against microbial infections. Lignin presents the same problem to bacteria as does cellulose as far as the initial attack upon it is concerned, and the sorts of organism that attack it solve the problem in the same way: they excrete an enzyme which partly breaks down the lignin outside their cells, and they absorb and consume the breakdown products. But the products of that breakdown can be toxic to most bacteria. That is why I deliberately used the word 'organism' just then. Though a few types of bacteria are able to attack and decompose lignin, resisting the toxic effect of the phenols so formed, they are not very good at it and the majority of microbes able to attack lignin are not bacteria but fungi. There are several kinds of 'wood-rotting' fungi, and some – such as the 'beefsteak fungus', which grows on senescent trees – are far from being microbes. Nevertheless, it is only representatives of the bacteria and fungi which have the ability to tackle lignin; higher organisms which consume wood, such as termites, rely on intestinal bacteria, some associated with protozoa, to conduct the process.

Another class of naturally occurring substances which only microbes consume is oil. Oil is a mixture of hydrocarbons – compounds of carbon and hydrogen only – and crude oil includes the volatile components which are distilled off industrially for use as petrol. Several kinds of bacteria, and some yeasts (which are micro-fungi), consume oil, oxidising it with air to obtain energy and to multiply. Though such microbes can be a nuisance where petroleum or oil is stored wet, they are very valuable in the aftermath of oil spills. For example, during the Gulf War of 1990–1, the Iraqi army deliberately released enormous quantities of Kuwaiti oil into the Persian Gulf, causing monumental pollution from which, some environmentalists feared, the Gulf would recover only slowly if at all. But they were wrong; earlier oil spills and seepages had prepared the microflora of its water for just such an event: within half a dozen months, by late 1991, microbial action in the relatively warm sea had rendered it cleaner than before the war started – cleaner because tankers had not returned in their pre-war numbers and there was less of the normally continuing oil pollution.

When oil spills, the microbes use the petroleum components most quickly; the thicker, less volatile material gets consumed more slowly, and a tarry residue, the stuff which ruins bathing beaches, persists longest – though it, too, disappears in the end.

Chemically, the simplest hydrocarbon of all is natural gas, known correctly as methane. It has appeared often in this book because many bacteria make it as part of their metabolism. The principal methane formers are a group of bacteria which flourish only in the absence of air, and therefore inhabit polluted sediments, sewage installations, compost heaps – and animal intestines, especially the stomachs of ruminants (they featured specially in Chapters 7 and 8). Huge deposits of methane exist under the earth's surface, often associated with oil deposits. Where it came from is still disputed; both bacterial and geochemical processes might have contributed. But the consequence is that methane is a normal, though minor, component of the atmosphere, seeping from these subterranean deposits, and being formed constantly by living things: it is a major component of marsh gas, and of the flatulence and belches of animals, including ourselves; it is even present in ordinary exhaled breath. In the air it is slowly oxidised by oxygen (except when it catches fire, when it is oxidised very rapidly indeed) to form water and carbon dioxide. Though it is an organic compound, it is useless to higher organisms. However, it is moderately soluble in water, and there it is oxidised by specialised bacteria which live on the top of sediments to catch the emerging marsh gas. (These methane-oxidising bacteria are completely different from the methane-forming bacteria of which I wrote a moment ago.) Similar bacteria inhabit the soil around natural gas vents. Methane molecules consist of one carbon atom surrounded by four atoms of hydrogen, and the bacteria use some of the molecules as their source of energy, oxidising them with the oxygen of air to carbon dioxide and water. With the energy so gained they use the remainder to make their own matter. Bacteria which use sugar burn about half of it and use the energy so gained to make their own matter out of the other half.

Another minor component of the atmosphere is carbon monoxide. This is the poisonous component of the exhaust gases produced by ordinary petroleum engines; it is very toxic to air-breathing things because it reacts with the haemoglobin of blood, destroying it, and it can also damage other vital parts of the respiratory system. But it is not only a man-made gas: it is formed naturally during combustion, and also as a minor product of bacterial metabolism in compost heaps and sediments. The air we breathe always contains small amounts of carbon monoxide, though it steadily, if gradually, becomes oxidised to carbon dioxide. Why are we not dead, you ask? Because we make more haemoglobin than we need, and can spare a little of it to remove such carbon monoxide as we are daily exposed to. Most people know that haemoglobin is essential for transporting oxygen about our bodies; few understand that it also functions as a normal detoxifying agent. Carbon monoxide poisoning results from overloading our haemoglobin. Once more, bacteria exist which exploit carbon monoxide, oxidising it to carbon dioxide and using the energy so obtained; curiously, they can also use gaseous hydrogen, another product of air-free microbial metabolism, as a source of energy.

Yet another poisonous natural product is hydrogen cyanide, a simple compound of hydrogen, carbon and nitrogen earlier known as prussic acid. Collectors know that if you crush a few laurel leaves in a jam jar, they will release enough cyanide to kill a butterfly. Cyanides also appear during the decomposition or combustion of nitrogenous organic material, including plant residues, though rarely in amounts sufficient to harm anyone. Nevertheless, types of bacteria, and some primitive fungi, are known which obtain energy for growth by oxidising cyanides; they have proved useful for treating industrial wastes that contain cyanides.

I could continue with a list of strange or toxic natural products which microbes of one kind or another (most often bacteria) can use as food, but it would become wearisome. Even seemingly stable carbon-containing matter is broken down slowly by microbes, such as coal, asphalt, peat, rubber and humic acids (the name given to the long-lasting organic substances that make soil brown). As a generalisation, it seems to be true that if a substance is of biological origin, either as a natural product itself, such as fluoracetate, or derived from one, like coal, then a microbiological process for consuming it exists. One might better ask, are there any organic substances which microbes cannot attack?

The answer is yes, but they are all man-made. Organic chemistry, the chemistry of the compounds formed by carbon, got its name for being the study of the compounds formed by organisms. But as soon as chemists – organisms themselves, of course – started manipulating those compounds in their laboratories, they made all sorts of substances which were hitherto unknown in nature. Today we are surrounded by them: familiar examples are plastics, synthetic detergents and artificial fibres; less obvious perhaps are explosives, refrigerants, dyes, pigments and some pharmaceuticals.

Some of these are indeed 'recalcitrant' in ecologists' language: there seem to be no ways in which microbes, or higher organisms, break them down. The refrigerants are causing anxiety at present, volatile compounds of carbon, fluorine and chlorine ('CFCs') which neither burn nor decompose biologically, so when they escape – or are released – they diffuse into the upper atmosphere, where ultraviolet light breaks up their molecules and the chlorine they release damages the ozone layer.

Plastics are another group of materials that often resist microbial attack. Some varieties can be consumed by bacteria, but the tougher kinds, such as 'PVC' and 'PTFE', are not. Ordinary sunlight decomposes most of them slowly, and some of the photo-decomposition products can then be tackled by bacteria.

A third group of problem substances comprises the so-called 'hard' detergents. A few years after detergents first came into general use, in the 1950s, scientists monitoring sewage and water installations realised that some kinds of detergent were not being removed by conventional sewage treatment: they were passing right through sewage works and emerging into rivers and lakes, ultimately creeping into drinking water. Sewage treatment relies wholly on microbes, which consume and clean up the human and industrial effluents which go to make up sewage. It became clear that the microbes could not handle certain classes of detergent – regrettably including some rather good ones – and in due course their use had to be restricted and ultimately discontinued. Perhaps the last residues of the heady days of detergents are still adsorbed in some aquatic sediment strata, for future geologists to discover!

Let me inflict one last group of recalcitrant substances upon you, before this list, too, gets tedious: the pesticides. The variety of chemicals which kill pests of one kind or another, and can properly be called pesticides, is enormous. They range from common salt (the scourge of snails and slugs) to complex synthetic organic chemicals such as 2,4-D (against weeds) and DDT (against insects). Salt washes away (much too soon to be of real use, except in a slug trap); 2,4-D is gradually destroyed by microbes and takes a few weeks to vanish from the garden; but DDT presents an environmental problem because it is very persistent. If microbes attack it at all, and there is no hard evidence that they do, they do it very slowly. That might not matter, because DDT is very active against insects and one does not need much, so you might expect the amounts that need to be spread into the natural environment would rapidly become so dilute, and also so much overlaid by natural deposits, that they ought not to matter. Unhappily, that is not so. After DDT had been widely used for 15 to 20 years, ecologists realised that in nature it underwent a process called 'bioaccumulation'. It gets eaten accidentally by herbivores, by scavenging worms, by insectivorous birds, fish and other animals, and its chemical nature causes it to accumulate in their body fat. It may do the primary consumers no immediate harm, but they have concentrated it. Then they are eaten, DDT and all, by other creatures, and the next generation accumulates more DDT. Though healthy adult higher organisms, other than insects, can generally take a moderate amount of DDT without obvious harm, there are limits, and some species are more sensitive than others. Soon whole populations of birds, amphibians and other creatures show signs of chronic poisoning. Most often it afflicts their eggs, larvae or offspring first. The chain of bioaccumulation of DDT is so well documented that in most temperate countries its use is now prohibited. But not everywhere, for what do you do, when abandoning DDT means that mosquitos return and tens of thousands of people die of malaria?

Bioaccumulation is not restricted to DDT; it can be a problem when other persistent chemicals get into a food chain: the substance may seem to be benign, except to the organisms under attack, but will later show toxic effects in completely unexpected ways.

Substitutes provide the only solution to the problem of recalcitrants. For the examples I mentioned above, 'ozone friendly' refrigerants which decompose at lower levels are coming into use (all too slowly, however); all household detergents today are 'biodegradable', which means that their molecules have been changed so that sewage bacteria destroy them; biodegradable plastics are being developed – one of the most 'environmentally friendly' is a rubbery material formed by a bacterium called *Alcaligenes eutrophicus*; and most herbicides and insecticides are biodegradable, though not always as rapidly as one would wish.

I must leave the economically important story of the environmental effects of recalcitrant substances because it is not really my theme here. For present purposes the surprising thing is not the failure of microbes to attack a few troublesome substances, but the remarkable fact that, of the enormous number of man-made organic chemicals that are in use for one reason or another, the vast majority *are* consumed by bacteria. Considering that the microbial world had not encountered these materials until twentieth-century man came along, how, and why, do they do it?.

The way fluoracetate works offers the first clue: the bacteria make mistakes. Organic chemicals actually fall into classes which resemble each other in chemical structure, such as phenols, sugars and so on. Therefore if a certain species of bacterium has evolved so as to be able to consume a plant phenol, for example, it might well consume a man-made phenol whose molecular structure is only a little different. Perhaps the man-made phenol will contain chlorine atoms, as a number of disinfectants do; the organism, or rather its enzymes, might not notice and, quite fortuitously, the chlorine atoms might be eliminated during the compound's metabolism, appearing in solution as harmless chloride. Or, as another example, the man-made substance might resemble a normal nutrient, such as a sugar, which bacteria use, and simply be attacked more slowly than the normal food. Perhaps its use benefits the organism, but instances are known in which substances are 'co-metabolised': they undergo the reactions of the regular substrate and are decomposed, but the organism.takes on the material by mistake, as a sort of metabolic passenger, and benefits not at all from its breakdown.

In effect, microbes can sometimes metabolise man-made products more or less by accident, using enzymes which ordinarily handle different, though chemically related, substances. This fact gives us a glimpse of how bacterial evolution has enabled some types to make use of quite alien materials. In the 1960s to 1980s, Patricia Clarke of University College, London, did an impressive series of experiments in which she 'taught' bacterial populations to consume new and unusual substances.

Professor Clarke was interested in the ability of a soil *Pseudomonas* called *P. aeruginosa* to use acetamide and related compounds (though not the fluoracetamide with which I opened this chapter). Acetamide is derived from acetic acid (the component of vinegar I mentioned earlier) but its molecule is modified to include a nitrogen atom, which effectively neutralises its acidity (it also causes it to smell uncommonly strongly of mice, a gratuitous piece of information that has nothing to do with my story). There are numerous compounds related to acetamide, collectively called amides. Clarke's bacteria consumed acetamide, and they would happily consume a closely related amide called propionamide (derived from propionic acid, a milk fermentation product which gives Emmental cheese its characteristic taste). But they could barely touch a somewhat more distantly related amide called butyramide. By compelling a population of her bacteria to make do with butyramide, she obtained a mutant which consumed butyramide many times faster than the original strain. It proved to have undergone a mutation in which the chemical structure of the enzyme with which the cell attacked acetamide (called 'amidase') had become altered. This mutant strain could still use acetamide. She then invited her new strain to use a substance called phenylacetamide, a substance which is also an amide but only remotely related to acetamide chemically. Like her original strain, her mutant strain could not do so, but eventually it yielded a mutant of itself which could. The amidase in this strain had changed yet again and, most significantly, it had lost the ability to attack its original substrate, acetamide.

That is a severely edited version of what Patricia Clarke did. Actually she conducted many more experiments, exposing populations of her *Pseudomonas* to various amides and elucidating the nature of the mutants she selected with considerable subtlety. I chose to describe that particular series because she forced her bacteria to evolve by at least two steps: first to thrive on butyramide, secondly to tackle phenylacetamide, and the second mutant lost the strain's original propensity to tackle acetamide. Thus it is a very clear indication of the way in which evolution, given appropriate natural selection, could cause a microbe to acquire ability to consume a wholly new substrate, while leaving no obvious clue as to where its new enzyme came from.

Other researchers working over the same period had compelled other bacteria, *Klebsiella* and *Escherichia coli*, to 'evolve' ability to attack new kinds of sugars. Like Clarke, they directed the evolution of their cells deliberately, and all relied on the appropriate mutations appearing spontaneously. That was a perfectly reasonable thing to do, because bacteria are very small and bacterial populations can often be huge: a billion cells in a thimbleful of well-polluted water is commonplace. Therefore spontaneous mutants are very common (this is an important thought, which I amplify in Chapter 17). So it is more than likely that the experiments really do mimic the way in which bacteria have learned to consume most of the man-made substances that they now encounter in nature.

It follows that all the man-made substances which bacteria attack are related in some chemical way to natural products that bacteria of one kind or another can consume. This may be the reason why DDT is recalcitrant; the structure of its molecule is in fact very unusual, in that it does not fit into any of the better-known groups of compounds that biochemists have recorded among the myriad of natural products; perhaps no organism possesses an appropriately amenable enzyme from which a 'DDTase' might evolve.

Evolution relies on mutations appearing spontaneously in populations, yielding mutant strains which are selected by the nature of the environment. Clarke and her fellow researchers also relied on mutations turning up spontaneously, though they imposed their own selection pressure. However, modern molecular biology has enabled scientists to move a step ahead of evolution, and to avoid the randomness of spontaneous mutation. Today, using the new technology of genetic engineering, it is possible to extract DNA from bacteria, to alter it chemically, and then to return it, altering the organism's hereditary characteristics. In effect, knowing what alterations might be necessary, the scientist is able to make the mutations he or she wants. Among features that molecular biologists can alter in this way are the structures of enzymes; they can change the rates at which they react with their substrate, and even make them prefer unfamiliar substrates. The possibility of creating wholly new enzymes to conduct reactions hitherto unknown in biology is being explored vigorously already. For example, one ought to be able to create a 'DDT-ase' – perhaps someone will have done so by the time you read this chapter.

Twentieth-century technology has imposed some rather nasty and persistent materials on this planet's inhabitants, and baffled our microbial cleaner-uppers. Perhaps we shall soon be able to provide our friendly neighbourhood bacteria with enzymes that evolution has not yet thought of, to help them clean up our mess properly again.

10

Of wraiths and ghosts

Unlike my earlier accounts, which have concerned strange and exotic habitats, conditions and diets enjoyed by microbes, this is a tale of beasties that emerge, wraith-like, from nothing, and of others which tempt and mislead the unwary, because they are not what they seem. But you may safely read my story late at night, because it will not give you nightmares (unless you happen to be a working microbiologist, when it might well). For the beasties are all microbes and none is harmful. And like all good ghost stories, my tale carries a stern moral.

Wraiths that live on nothing

I encountered a distinctly mysterious situation around the turn of the 1960s, when I was studying what happens to vegetative bacteria if they are starved. The simple answer is that they die (more about that in Chapter 16). But my interest was a mite more subtle: I wanted to know how fast and why. As part of my research plan, I repeated experiments done by Arthur Harrison, then of Vanderbilt University, USA, who had discovered a rather peculiar effect: that dense populations survive starvation for longer than sparse ones: within limits, crowded bacteria seem to help rather than hinder each other's survival. I must not digress into why this happens; my particular experiment was to test Harrison's 'population effect' with my bacteria. My organism, the same as Harrison's, was *Klebsiella aerogenes*; I had chosen it because it was easy to culture, it needed no vitamins or complex organic food supplements: it grew readily with just glycerol and a few mineral salts, dissolved in water. And I had already worked out how to measure its death rates quickly and accurately.

My method was to spread tiny samples of starving culture, diluted if necessary, on thin films of nutrient jelly on a microscope slide, and after a few hours in the warm I would look at the slide under the microscope and count what proportion of the cells I had put there had formed micro-colonies and what proportion had not. Those that had not were dead, so I could easily work out what the ratio of dead to live cells had been in the original sample. (I must add that I had two accomplished helpers, Janet Crumpton and John Hunter, to help me, otherwise I'd have gone barmy, incessantly counting small objects. Even so, I think we all three got a bit light-headed after a heavy day's counting.) Anyway, I set up populations ranging from some 100 million to a hundred in a drop, and left them to starve in my routine conditions (aerated and warm, in a dilute solution of salts with no food), and examined samples of the populations at intervals.

Most of the experiment went according to expectation: populations from 100 million to about a thousand per drop died at increasing rates. But imagine our surprise when the very sparse populations, about 100 to 1000 per drop, remained doggedly alive. A clue arose when, after a few hours, one of us noticed that there seemed to be more colonies on our counting slides than cells originally put there: sure enough, the sparse populations were becoming less sparse. The damn things were multiplying.

This seemed ridiculous, but it happened every time. Well-trained microbiologists that we were, we had gone to tremendous lengths to exclude unwanted nutrients from our solutions, especially organic matter which would permit our bacteria to multiply. We were also very careful about the way we handled our materials, solutions and apparatus, and at pains to ensure that our bacterial populations did not carry extraneous organic food into their starvation vessels.

Yet, wraith-like, new bacteria seemed to be emerging from nothing. Nonsense, of course. Somehow they were finding enough food in our starvation media to enable them to multiply until they reached about 1000 per drop. This was far too few to make the liquid cloudy; it was 'gin clear' as my then superior, Denis Herbert (a noted biochemist), put it, and for almost every ordinary microbiological purpose it would have passed unnoticed.

A tedious series of experiments told us that, despite our using the purest grades available, every chemical that we used in our starvation media contained minute traces of organic matter which the bacteria could use. So did our doubly-distilled water. The numbers of bacteria that appeared told us that the levels of contamination were well below the purity limits guaranteed by the manufacturers: we could not complain.

We redesigned our experiments to get round our particular problem. Over the next few months, other microbiologists in the building also found 'growth on nothing' in seemingly non-nutrient mineral solutions. A common example would be a solution of potassium phosphate in a bottle on a shelf, in which a sparse, filamentous clump appeared slowly over the weeks, looking just as if a tuft of cotton wool had dropped in and swelled out. Several of our colleagues thought at first that it was just that; in fact, under the microscope it proved to be a micro-fungus, like a bread-mould, making do with consumable impurities of some kind. And soon we learned that other laboratories often experienced, and almost as often disregarded, microbial inhabitants in their stock solutions.

Well, impurities in commercial reagents provided a partial explanation of our wraiths. In fact, such impurities are no surprise to most microbiologists: one is warned against them in one's training. But organic nutrients in doubly-distilled water was another thing. How could it come about?

Whatever was there was far too dilute to be detected by ordinary analysis. Evaporating down a lot of distilled water left nothing visible, so whatever-it-was was volatile. And that gave us our clue: it was something in the air.

We never found out exactly what it was, but it happens in every laboratory. Alcohol, for example, is used all the time in laboratories: as a sterilant, to wash glassware so that it dries quickly, as a solvent to clean greasy apparatus, as a reagent, as a component of media for microbes. It is present in the air, though in minute concentrations, and it dissolves in water. When I looked into this matter I came across a remarkable photograph of a slowly dripping tap investigated by scientists at London's Metropolitan Water Board. It had a long stalactite of mucilage hanging from it, rich in bacteria, yet there was not nearly enough organic matter in the tap water to allow such abundant growth. The explanation proved to be that there had been for months a large dish of industrial alcohol beside the sink, into which contaminated glassware was dropped, to be cleaned properly at the end of the day. Alcohol vapour had been steadily diffusing across to the tap, dissolving in each drop of water, and enabling the bacteria to multiply. Their ability to form mucilage ensured that they were not washed away: they adhered to each other, and to the orifice of the tap, in a glutinous mass.

Another source of volatile bacterial food is human sweat. The 'stale' smell of sweat is compounded in part of ethyl and butyl alcohols, acetone and chemicals called isoprene and toluene; the first three can be 'eaten' by many kinds of bacteria, as well as by some micro-fungi, and all three dissolve readily in water. Exhaled breath can contain several of these substances, too. They are present in the atmosphere wherever people are around and, where people are crowded, they can cause serious problems. Art galleries and museums have to have good air-conditioning systems to protect the exhibits from the chemicals which condense out of their clients' sweat and exhalations, as much as from pollution from the traffic in the streets outside. Prehistoric paintings discovered in caves at Lascaux in France in the 1960s had to be closed to the general public because the volatile emissions of visitors were enabling microbes to grow on and spoil them; when in 1980 I had the good fortune to visit the ancient tombs at the Valley of The Kings at Karnak in Egypt I became convinced that something of the sort was happening there; I believe remedial action is in hand.

Microbiologists who are aware of this sort of thing, and not all are, now recognise a class of bacteria which specialise in living on very dilute substrates. (They are called by the formidable name of 'oligotrophs' and some are so used to living in nutrient-deficient conditions that they actively dislike the rich media commonly used by microbiologists, so they can be difficult to cultivate in the laboratory.) They are found in clear springs and lakes, for example. But our *Klebsiella aerogenes* was not of this class, it loved lots of food, but it was still able to emulate oligotrophic growth.

The phantom fixer

Perhaps the weirdest example of nutrient scavenging by microbes is one that I tangled with a few years later, one which has caused on-going confusion in the area of biological nitrogen fixation for many decades.

I shall write of biological nitrogen fixation, and its peculiar enzyme, in Chapter 11, but I shall preview the relevant bits here. Briefly, it is the process whereby nitrogen from the atmosphere is converted into a form which plants can use for growth, thereby supplying the major food sources of all living things. An underlying principle of the subject is that only bacteria are able to fix atmospheric nitrogen. No higher organisms seem able to do it. Nor, indeed, can most bacteria: the property is confined to about 100 out of the 10,000 or so bacterial species named. So-called nitrogen-fixing plants are not nitrogen fixers themselves: they are conventional plants that enter into symbiosis with one or another species of nitrogen-fixing bacteria. Perhaps the best known are the legumes (peas, beans, clover and so on), which associate with the genus *Rhizobium*.

That is the rule: only bacteria fix.

Microbiologists, like all good scientists, dislike such rules and would be happy to see it undermined but, despite numerous attempts, it has so far stood the test of time.

Yet for some 150 years spectres have haunted the subject. Phantom nitrogen fixers, only rarely bacteria, have lurked in strange places. A beguiling apparition was described in a letter to *Biologist*, the magazine of the Institute of Biology, in June 1987. G. B. Firth, a chemist of St. George's School in Rome, wrote that in 1986 he noticed an insect larva in his sugar bowl. With commendable scientific acumen, he segregated it in a jam jar for further observation, together with some of the sugar for sustenance. About 12 months later the larva had cocooned, and after about 5 more months it duly metamorphosed into a small but active moth. Where, asked Firth, did it get the necessary nutrients, other than the carbon, oxygen and hydrogen of sugar, to grow and to metamorphose? Could he be the first to discover a nitrogen-fixing insect?

I think not, and shall return to the matter later.

Actually, this was not the first time invertebrates had been suspected of fixing nitrogen. In the 1960s, John Millbank of Imperial College, London, and Michael Baker of the Forest Products Research Laboratory (now Forest Research Station) in Surrey, looked at the matter of the common woodworm, *Anobium punctatum*, a beetle that bores into wooden furniture, beams and so on. This insect lives on a diet of wood, a carbohydrate material containing virtually no combined nitrogen, so how does it take up enough nitrogen to grow? Could it be that it fixes nitrogen? Or perhaps that, like some plants, it carries symbiotic nitrogen-fixing bacteria? By the 1960s new techniques, microbiological and biochemical, were available to test both of these propositions, but Millbank could find no evidence to support either.

Termites, too, live on wood and face the same problem. In their case, scientists at two laboratories in the USA showed in the early 1970s that, though termites do not themselves fix, they do have nitrogen-fixing bacteria in their guts. But so do people, cattle, pigs and doubtless most animals; in these species we know that, although present, the bacteria do not fix a significant amount of nitrogen. For example, in 1973 Peter Hobson, of the Rowett Research Institute in Scotland, and David Ware and I, working at Sussex, measured rates of fixation in the rumen of a sheep starved of fixed nitrogen: the rate worked out at about 0.4 milligrammes fixed per day. That was, as I recall, less than 1% of a healthy sheep's normal requirement. Comparable research by American workers on termites suggested that they did a bit better: 2 to 3% of their nitrogen intake could come from nitrogen fixers in their guts. There has been a suggestion that certain inhabitants of Papua, New Guinea, who live on a very restricted diet containing extremely small amounts of nitrogen, benefit from the nitrogen-fixing bacteria in their intestines – but again the sums do not work out.

Where researchers have measured the rates of bacterial fixation in these potential symbioses, they have turned out to be trivially small. The current record holder is the shipworm, an aquatic mollusc which bores into the timbers of ships. This creature has symbiotic cellulose-decomposing, nitrogen-fixing bacteria in a gland off its intestinal tract. According to scientists at Harvard University and Woods Hole Marine Biological Laboratory in Massachusetts, USA, these can contribute as much as a third of the molluscs' nitrogen requirement, though the molluscs look pretty sick if this is their only source of nitrogen.

Thus it seems that, though nitrogen-fixing bacteria do associate with some animals, they generally provide their hosts with very little nitrogen indeed. So where does the rest come from?

Some scientists have made extravagant claims on behalf of higher organisms, suggesting that they fix nitrogen without bacterial aid. In the 1960s, E. M Volski, a scientist from the (then) Soviet Union, achieved a kind of notoriety with a series of papers and a monograph that maintained that all sorts of biological systems fix nitrogen, including bee pupae, silkworm chrysalises, hen's eggs and even quail embryos. Eventually, in 1970, a dozen Soviet Academicians published a thunderous refutation in one of the USSR Academy of Science's journals. Earlier, in 1963, one F. V. Turchin and his colleagues, also in the Soviet Union, had presented evidence that nitrogen-fixing enzymes are present in the stems and leaves of all sorts of plants, including legumes and fungi. No-one has positively refuted Turchin's claim but it has become less and less plausible as scientists have begun to understand the peculiar biochemistry and genetics of the enzyme nitrogenase.

In 1980, two scientists from Tokyo achieved transient fame by publishing a plausible paper describing nitrogen fixation by a culture of a green alga, a species of *Chlorella* that lives in hot springs. This was remarkable because that organism is a true plant, albeit primitive. But no further reports appeared, and when John Millbank checked a subculture of the alga he found no evidence of fixation. In another test, I checked some taxonomically similar strains, isolated elsewhere, and again the results were negative. Yet *Chlorella*'s name reappeared in this context in 1984, this time from a group of scientists working at Worcester Polytechnic, Massachusetts. Again, the evidence for nitrogen fixation was not definitive and no further substantiation appeared.

Yeasts, like other fungi, are often regarded as true plants and the scientific literature includes many reports of nitrogen-fixing yeast cultures, several of them quite convincing. But Millbank – an indefatigable layer of yeast ghosts – obtained negative results with every such culture that he tested. I checked a couple myself, and also an *Aspergillus* (a filamentous fungus) from Poland, with similarly negative results. No-one, to my knowledge, has checked several reports of nitrogen-fixing mushrooms, a recent example being a species of *Pleurotus* reported from India in 1980. In these cases, however, there is not enough evidence to show that associated nitrogen-fixing bacteria were absent, so the claims remain dubious.

One of the more exotic phantoms was raised by A. H. Laurie in 1933. On an expedition in the research ship *Discovery*, Laurie was struck by the fact that whales dive quickly to enormous depths and, after a period, surface just as rapidly. A human would suffer – probably die – from the bends after such a rapid sequence of changes in pressure. On the downward plunge compressed nitrogen would dissolve in the blood; on the upward run the nitrogen would be released in the form of bubbles, with their attendant risk of embolism.

Why, Laurie asked, do whales not get the bends? His solution, an intriguing idea at the time, was that whales have nitrogen-fixing bacteria in their blood. The bacteria, he argued, fix nitrogen so fast that it does not re-emerge as gas. Laurie actually presented reasonable-looking evidence for nitrogen-fixing bacteria in the blood of recently captured whales.

We now know that even if the blood were a thick bacterial soup and the bacteria fixed at the fastest possible rate, they could never mop up dissolved nitrogen fast enough. In 1949, J. Case showed that Laurie's nitrogen fixers, which were real, were *post mortem* contaminants from outside the whales' blood system. However, Laurie's question was a serious and interesting one. I learned in the mid 1960s from a zoologist that, although a whale's lungs can indeed compress to the size of a couple of purses at great depths, the animal has such a large volume of blood that it can dissolve, and later release, all the nitrogen without bubbles forming.

Exit one ghost. But the others remain; and there are yet more, even among the bacteria. Sometimes bacterial ghosts – bacteria which were thought to fix nitrogen but did not really do so – proved to be a matter of poor microbiological technique. For example, a few reports of nitrogen fixation by ordinary soil bacteria appeared because researchers did not realise that they had failed to obtain a pure population and were unwittingly studying a mixture of two types, one of which did not fix, the other of which did. However, *Pseudomonas azotocolligans, Azotomonas insolita* and *Nocardia cellulans* are the names of cultures of bacteria deposited in official Culture Collections as nitrogen fixers which, on close examination with more modern methods, have proved not to fix nitrogen. Yet they were authentically pure cultures; what had gone wrong?

Once I isolated a couple of ghosts myself. It was around 1964, when I first started research on nitrogen fixation, and at the time I did not think they were ghosts. It was the laying of my ghosts by Susan Hill and me which gave us the clue to what was happening elsewhere. I had isolated my strains of bacteria while I was working for a few months at the Royal Veterinary College in Camden Town, London. They were undoubtedly pure cultures; they were rod-shaped bacteria and they grew nicely on the surface of jellified, nitrogen-free culture media. However, circumstances caused me to move my research to the University of Sussex, where my organisms would grow only very, very slowly, unless I gave them a little fixed nitrogen, as an ammonium salt.

In due course light dawned: at the Royal Veterinary College I had worked in a laboratory five floors up from the stables, which housed animals. From these we occasionally caught a whiff of ammonia, wafting gently upwards. At Sussex there was no such thing. Ammonia is a form of fixed nitrogen, and my isolates, as well as the 'official' nitrogen fixers when we came to check on those, proved to be remarkably effective scavengers of fixed nitrogen gases, including ammonia, from the atmosphere. They not only scavenged such nitrogen, they made do with very little. About 14% of the dry weight of ordinary bacteria is nitrogen; our ghosts made do with 3 to 6%. Why did the Culture Collection not spot this when it tested its own cultures of the bacteria deposited as fixers? Because it was located at Aberdeen, across the estuary from the ammoniacal effluents of that city's famous fish market. Instigated by us, they soon satisfied themselves that the named strains were ghosts.

Where does the fixed nitrogen in the air come from? Many dust particles are rich in all sorts of forms of fixed nitrogen, coming from dried body fluids, airborne microbes, fragments of dead insects, tiny flakes of skin, dry mucus from sneezes or nose-blowings, and so on. But cultures of microbes are largely free of dust. (Some scientists fondly believe that dust is excluded completely, but they forget that dust gets into the reagents and water used to make up microbiological media, and that the most common gelling agent, agar, always has dead bacteria in it, serving as a source of nitrogen.) In a laboratory, nitrogen may also come from reagents such as ammonia or nitric acid (we had banned ammonia from our Sussex laboratory).

In a well-run laboratory these sources of fixed nitrogen do not amount to much. But even when banned, ammonia is ubiquitous. Apart from obvious sources – stables, manure heaps, babies whose nappies need changing – ammonia is present wherever there are people, coming from their breath and sweat, along with the organic nutrients I mentioned earlier. Ammonia is also present in tobacco smoke, happily a much rarer hazard in laboratories than it once was, and it is also a component of some proprietary cleaning agents and polishes which are widely used by institutional cleaners. It is normally a dilute component of the atmosphere, usually well below the threshold at which one can smell it, but a bacterial colony on the surface of a jellified medium in air, low in nitrogen, will consume all it can get. And as the concentration of ammonia in the air close to the colony becomes vanishingly small, more ammonia diffuses towards the colony to replace it. In effect, as they consume even traces of ammonia, the bacteria set up what is called a diffusion gradient, operating in their own favour.

Diffusion gradients can be important in many aspects of the nutrition of microbes. Whenever a microbe is consuming a gaseous, or water-soluble, nutrient, it will scavenge that substance from its immediate vicinity and more will move in to replace it by the natural processes of diffusion. It is called a 'gradient' because the concentration of the nutrient gradually declines the closer you get to the cell and becomes none-at-all within the cell. Diffusion gradients occur in non-living systems, too. A solution of alkali open to air will steadily pick up carbon dioxide; gas masks rely on the cumulative effect of millions of tiny diffusion gradients round particles of adsorbent within the mask, which collectively lower the concentration of the noxious gas to zero.

To return to microbes. Another source of fixed nitrogen is any machinery that involves either electric sparking or an internal combustion engine – including lorries delivering supplies near a ventilation intake. These generate small but steady amounts of nitrogen oxides which, forming nitrates when they come into contact with water, can be utilised by most types of bacteria, who thus set up a comparable diffusion gradient.

It is not easy to clean all the fixed nitrogen contaminants out of air; you need to 'scrub' it through a filter capable of removing dust, through alkali to remove the nitrogen oxides and then through sulphuric acid to remove the ammonia. Even with these precautions a little fixed nitrogen always finds it way through. The most remarkable ghost of my experience was brought to our laboratory in the early 1970s by a visiting scientist from Canada, who was sure he had a nitrogen-fixing yeast (a pretty pink *Rhodotorula*). It grew in 'scrubbed' air, though slowly. But none of its other properties was consistent with nitrogen fixation. The upshot was that *Rhodotorula* was a remarkable scavenger: in these conditions its cell composition was only 1% nitrogen, yet it continued to grow.

I have wandered rather far from Mr Firth's nitrogen-fixing moth, but one can see now that there are two plausible explanations that do not require it to be a nitrogen fixer. One is that the sugar was less free of fixed nitrogen than he thought; sugar is a plant product, after all, which must carry traces of nitrogenous compounds through the refining process as well as picking some up from dust. But the larva would have to consume rather a lot of sugar to keep a proper balance of nitrogen. I strongly suspect that Firth's moth had in its gut microbes that are good at scavenging fixed nitrogen from the atmosphere. I wonder how far St. George's School is placed with respect to the nitrogenous effluents arising from the citizens of Rome, and from their abundant, petrol-burning traffic?

As to the other phantoms, the *Chlorella* and yeast ghosts probably arose because those organisms possess the necessary enzymes to scavenge both ammonia and nitrogen oxides. Careful reading of the 1984 paper on *Chlorella* reveals that the authors underestimated the potential contribution of atmospheric nitrogen oxides. But one must attribute other reports, such as fixation by higher fungi and the 1980 report on *Chlorella*, to unrecognised nitrogen-fixing bacteria lurking in the system somewhere. In sheep, cattle and so on I incline to the view that the nitrogen fixers one regularly finds there are passengers: nitrogen fixers from soil and plant material which they take in with their food.

There is an ironic historical twist to the story of the phantom nitrogen fixers. The discovery of nitrogen fixation is usually assigned to two German scientists, Hermann Hellreigel and Hermann Wilfarth. From 1886 to 1888 they published definitive evidence for fixation by legumes and demonstrated their need for the bacteria which live in leguminous root nodules. In reality, the process was discovered and reported fifty years earlier by a Frenchman called Jean-Baptiste Boussingault. The irony is that

Boussingault did not accept his own results: today one can see from his data on the nitrogen contents of experimental plants that he had pea plants, which fixed nitrogen, and oat and wheat plants, which did not. Boussingault attributed the differences in nitrogen content to the legume's superior ability to scavenge atmospheric ammonia and nitrogenous matter in dust – to form steeper diffusion gradients, though he did not use the term. He thought his legumes were what we now call ghosts, and this held up the discovery of biological nitrogen fixation for more than half a century.

The moral

Yes, I promised you a moral. It is obvious. Read, mark, learn, inwardly digest – and then doubt what they tell you. It is, indeed, a message cryptically enshrined in the motto of the Royal Society: *nullius in verba*.*

---

* A contraction of a line from Horace: *nullius addictus iurare in verba magistri* (in the word of no master am I bound to believe).

11

The inertness of nitrogen

The elements

For millennia, philosophers have wondered about the nature of matter, about what the solids, liquids and gases which compose and surround us are made of. Aristotle asserted that material substances consisted of air, earth, fire and water in different proportions, and this view, at least in principle, held sway until the late seventeenth century, when Robert Boyle initiated modern understanding of the real nature of the chemical elements.

Early in the nineteenth century, John Dalton, a father figure of modern chemical theory, proposed that the elements consisted of unique, identical atoms, an advance in understanding that gave the subject tremendous impetus. But almost a century later chemists learned that this was not strictly true: the element nitrogen, for example, consists mainly of one kind of atom, but a minority of slightly heavier atoms is always present, both in the native element and in its chemical compounds. The heavy atoms have chemical properties so nearly identical to those of the lighter ones that the two kinds of atom, called isotopes, do not become separated during chemical transformations. So how does anyone know they exist, you ask? The answer is that one can distinguish them by physical means; for example, by causing them – or a compound containing them – to sediment in a powerful gravitational field, when the heavier one sinks the faster. Most elements are mixtures of isotopes; nitrogen has only two, but some have several. These days isotopes can be separated routinely – as witness the atom bomb, which is made from a scarce isotope of the metal uranium which has been separated out from the natural mixture – and today one can buy chemicals enriched with certain isotopes, or even containing exclusively one isotope, for use in research and medicine.

When I was a boy in the 1930s I was taught that there were 92 chemical elements. Only 90 elements, my teacher added, had actually been discovered; a precept that promptly aroused my suspicion – how could anyone know of the existence of an undiscovered element? But I learned quite soon that this belief was based in sound theory about the nature of atoms, and I was duly gratified when the missing pair were discovered several years later, by which time I was a young scientist myself. Today upwards of 102 elements are known, most of the newer ones having been found among either the products of radioactive decay, or created by physicists in monstrous atomic particle accelerators. They are not part of our everyday world.

In fact, as far as daily life is concerned, we encounter only about thirty of the chemical elements regularly, and they are usually present in chemical combination with one another. We rarely see free elements. Consider the few examples that we do see. Elemental carbon is common enough as coal and soot; it also turns up, rarely, as diamond. The element sulphur appears in sulphur springs, sometimes forms a whitish scum on polluted pond water, and can be mined in special localities. Gold is found, very rarely, in them thar hills; so, even more rarely, are a few precious metals such as platinum and palladium. As far as uncombined elements visible in the inhabitable parts of this planet go, that is about all, though there would seem to be a lot of elemental iron at the Earth's core, with perhaps some other free elements in or around it. Visible elements at the Earth's surface are outnumbered by nine invisible ones: those that are gases in ordinary conditions. Oxygen and nitrogen, the major components of the atmosphere, are uncombined elements, so are the 'rare gases' (helium, neon, argon, and three more) which form a small proportion of air. And there is always a little elemental hydrogen in the atmosphere.

What determines whether an element occurs uncombined in nature? Some, the rare gases, are there as elements because they simply do not form stable compounds with other elements. Others, such as gold and platinum, do form compounds, but with difficulty, and their compounds break down easily to release the free element. But hydrogen, oxygen and sulphur are very reactive; they very easily form stable compounds with other elements, and the only reason why they turn up uncombined at all is because living things are constantly regenerating them. Oxygen, for example, is too reactive to last long on a lifeless planet, but it is released from water by plants during photosynthesis; elemental sulphur deposits are mainly formed by certain types of sulphur bacteria (I wrote of them in Chapter 7); hydrogen is generated by putrefying bacteria as they decompose organic matter – it is belched out by cattle, too, and is a component of fart. Such free elements are passing through the environment, so to speak, generated by biological processes and mostly destined to re-enter chemical combination, sometimes spontaneously, often by way of yet other biological processes.

Elemental nitrogen is the odd-ball. It does not enter into chemical combination at all readily and the free element is abundant. Nearly 80% of the atmosphere is free gaseous nitrogen, amounting to an incredible 3.9 million billion (3×10^{15}) tonnes of gas. Yet most of its compounds, once formed, are quite stable, and every living thing, plant, animal or microbe, contains compounds of nitrogen, such as proteins, nucleic acids, hormones and vitamins. But, with a tiny group of exceptions which I shall come to shortly, none can get its nitrogen atoms from the elemental gas. The raw material for these nitrogenous materials comes from food, and that food comes ultimately from plants (and a few plant-like microbes, such as yeast in bread). These make their nitrogen compounds from nitrates, or occasionally ammonia: simple compounds of nitrogen which they find in fertile soil. The nitrogen compounds in living things are recycled: animals eat plants and use their nitrogen compounds; we eat both animals and plants. Excretion, death and decay pass nitrogenous compounds back to the environment, where bacteria convert them to nitrates for plants to re-use. This brisk turnover works well, but the cycle leaks: certain kinds of bacteria exist which convert nitrates to nitrogen gas, and this escapes to the atmosphere; then neither plants nor animals, nor yet the majority of microbes, can make use of it. It returns to being an inert diluent of the oxygen all these creatures respire.

If that were the whole story, life on this planet would not have got very far over the aeons: evolution would have been gravely limited by declining supplies of recycled nitrogenous compounds on the Earth's surface. These would have been supplemented only modestly by the small amounts which are formed chemically from nitrogen gas, in flames and lightning discharges for example, plus a little that appears in volcanic emissions. However, for at least a third of life's lifetime, so to speak, a rare class of microbes, called the nitrogen-fixing bacteria, has been topping up the nitrogen cycle, off-setting the leakage by bringing new nitrogen back from the atmosphere into biological food chains. The bacteria have been able to do this because they possess an enzyme called nitrogenase; and only in the last hundred years has mankind begun to supplement the bacterial input, by spreading nitrogen fertiliser made industrially from the atmosphere by the Haber process. (At the end of the twentieth century, Haber nitrogen fertiliser accounts for almost a quarter of the world's input of newly-fixed nitrogen compounds; nitrogen-fixing bacteria provide just under two thirds and the rest comes from lightning, combustion and minor non-biological sources.)

The how . . .

Nitrogenase, then, is the enzyme responsible for nitrogen fixation, the name given to the conversion of nitrogen from the atmosphere into a chemical form which plants can use for growth. Along with the primary enzymes of plant photosynthesis, nitrogenase is an enzyme of transcendent importance to all living things. Yet it is absent from the vast majority of them; it is found only among bacteria, and then only in less than a hundred of the thousands of species known. That is one reason why it is fascinating to biologists: why is its distribution so restricted? I shall return to that question later; first I shall consider an equally curious chemical puzzle. The reaction that nitrogenase catalyses is the formation of ammonia from atmospheric nitrogen, a reaction which is difficult to perform in the laboratory, especially when water and oxygen are around. Because nitrogen is a remarkably stable gas, rather violent conditions are needed to make it enter into chemical combination. The Haber process, for example, requires dry, oxygen-free gases (hydrogen and nitrogen) to be brought to high pressures and temperatures and exposed to specialised metallic catalysts, before they combine to form some ammonia. The vexing question is, how do certain bacteria pull off this chemical task in air and water at ordinary temperatures?

It seems that it is not easy. Indeed, if there were a *Guinness Book Of Records* for awkward, complicated enzymes, I am sure that nitrogenase would be among the front runners. It has proved to be one of the most difficult enzymes to isolate and characterise; even today, over thirty-three years since the first crude extracts containing it were prepared, it is awkward to handle and study, and it is highly unlikely that a pure, wholly undamaged preparation of nitrogenase has ever been obtained. And those properties that have been revealed, and they are many, seem to be almost wilfully perverse.

A major oddity of nitrogenase is that it consumes energy. For every molecule of nitrogen gas that is reduced to ammonia, sixteen molecules of ATP are expended. (I wrote about ATP in Chapter 7, but I remind you now that it is the biological energy currency: growth, movement, maintenance and all the other things cells do use up ATP; food and respiration together generate ATP.) The ATP has to be provided by the organism – or by the biochemist – or the enzyme does not work. The Haber process consumes energy, too, but in fact there is no theoretical reason why the enzyme should do so. If you calculate the energy balance of an *Azotobacter* oxidizing glucose and reducing nitrogen gas, which is easily done, it comes out favourably: the arithmetic says that the organism ought to gain a little energy over all. The need for ATP remains a vexing question.

I shall not enlarge on the further fact that if you give nitrogenase too much ATP at a time it rapidly stops working. That is a minor nuisance, one which delayed the discovery of the ATP requirement for a mere two years. A more blatant perversity is the fact that nitrogenase is destroyed by air. Contact with oxygen inactivates nitrogenase within seconds and it cannot be reactivated. The remarkable thing is that this happens even if the enzyme is purified from, for example, *Azotobacter*, a bacterium which not only grows, and fixes nitrogen, in air, but will not do so without oxygen. The enzyme's sensitivity to oxygen is very inconvenient for biochemists, who have had to develop a whole new technology for handling air-sensitive proteins in order to purify and study nitrogenase at all. And, if I may digress a moment, it is also very inconvenient for the bacteria themselves, who have to go to considerable lengths to keep oxygen out of the way when they fix nitrogen. That is another matter that I shall return to.

The reasons for this oxygen sensitivity are still a mystery. One could understand that oxygen might interfere with the enzyme's working: basic chemistry suggests that any enzyme that could bind and activate nitrogen molecules might well pick up and react with oxygen molecules by mistake, as it were, but it happens also to the passive, non-functioning enzyme, when it has no ATP. There is no clear reason why it should be destroyed by oxygen at all, let alone so rapidly.

And there is another awkwardness to reveal. I have written of nitrogenase as an enzyme, but the singular noun is not correct. It is a complex of two distinct enzymes, which can be separated.

One is a larger and more complex protein than the other; both are intolerant of oxygen and both are essential for activity. So which is the real nitrogenase? It took over a decade to sort them out; we now know that the larger protein binds the nitrogen molecule, but it will not do so unless it is also reacting with the smaller in a way that enables it (the larger one) to generate from water some hydrogen, which it needs to convert the nitrogen to ammonia. In this two-fold reaction, the smaller protein consumes ATP. This does not really explain why nitrogenase needs ATP, but it does suggest at what stage it needs it.

I said that the larger enzyme, on reacting with the smaller enzyme, together with some ATP, generates hydrogen from water. Indeed it does. If there is no nitrogen gas around, as when a researcher sets some nitrogenase to work under an inert gas such as argon, it just blows off the hydrogen, which it can no longer use. But the silly thing is that it releases hydrogen even when nitrogen gas is present: for every molecule of nitrogen converted to two of ammonia, one molecule of hydrogen is formed. This formation of hydrogen represents a waste of ATP, and it happens in real life, as well as in a laboratory: Harold Evans and his colleagues, of Oregon State University, USA, showed more than a decade ago that, over a field of soya beans, there is a gentle waft of hydrogen into the air, coming from the nitrogen-fixing bacteria in the bean plants' root nodules. The more efficient kinds of nitrogen-fixing bacteria, such as *Azotobacter*, have an enzyme system which picks up the hydrogen before it escapes, and can use it to conserve ATP.

But why should hydrogen be formed at all? The answer seems to lie in a detail of how nitrogenase works. The molecules of both component proteins contain atoms of the metal iron, but the big one contains molybdenum atoms as well; two in all. And evidence has accumulated steadily over the last 15 years that the molybdenum atoms are either the actual sites at which the nitrogen gas is bound and converted to ammonia, or else they are closely involved with that site. Chemists and biochemists imagined for a decade or more that the involvement of molybdenum went something like this: nitrogenase has first to form a molybdenum hydride, as it is called, and the hydrogen of the hydride group then becomes displaced as nitrogen is bound, and is released as free hydrogen; then the bound nitrogen picks up more hydrogen atoms from water and, aided by ATP, becomes ammonia, which is a nitrogen hydride. Such reactions can be mimicked in the test tube. Complex molybdenum compounds have been made by chemists which will convert nitrogen molecules, bound to the molybdenum atom, to ammonia in watery solutions (elemental oxygen must be absent), and others which show displacement of hydrogen from a molybdenum hydride by nitrogen gas.

Nitrogenase is not very good at recognising the nitrogen molecule; it mistakes a number of small molecules, such as acetylene, for nitrogen and adds hydrogen to them instead. But this, too, is the sort of thing a metal atom such as molybdenum might reasonably do when bound into a protein, and the evidence for such a mechanism is now quite good. Yet it seems a laborious way of doing things. And so it proves to be: nitrogenase is one of the slowest-working enzymes known to biochemists. The consequence is that, in order to fix reasonable amounts of nitrogen, bacteria have to make an awful lot of it; nitrogenase amounting to 10% of the microbe's protein is common, and up to 40% has been recorded. Simply making enough enzyme must be a substantial burden for a growing cell. And all this generates another query: nitrogenase working within the cell must obviously be highly concentrated, but nitrogenase being tested by a biochemist in a laboratory is necessarily more dilute by orders of magnitude. How relevant are laboratory results to real life, then? We do not know. There are reasons for suspecting that the problem of obtaining really undamaged nitrogenase proteins in the laboratory arises at least in part because the proteins' structures become disturbed during dilution. But despite all these reservations, a lot has been learned about how nitrogenase works.

For a dozen or more years the concept of a central role for molybdenum in the working of nitrogenase went almost unquestioned among scientists. Then, in the mid 1980s, Paul Bishop of North Carolina State University, USA, used the techniques of bacterial genetics to show that a species of *Azotobacter* could, if necessary, make nitrogenase without molybdenum at all. However, this was not quite the blow to theory that it might have been; my erstwhile colleagues at the University of Sussex, Robert Eady and Robert Robson, showed that a comparable aberrant nitrogenase, albeit made by a different species of *Azotobacter*, contained atoms of the metal vanadium instead of molybdenum. This made reasonable sense chemically: vanadium atoms can bind nitrogen gas in appropriate chemical environments, and might behave very like molybdenum as part of a metallo-protein molecule.

Calm and enlightenment were short-lived. Within a dozen or so months, Bishop's species of *Azotobacter* was shown to be able to form a third kind of nitrogenase, one which contained neither molybdenum nor vanadium. The only metal atoms in its structure appeared to be of iron. And there was no question of all these new nitrogenases being laboratory artefacts of some kind: the microbe has special genes coding for this new enzyme, as well as genes coding for the vanadium enzyme; these have been identified – even isolated as DNA clones – and they are not the same as a third set which codes for its molybdenum nitrogenase.

So, during the 1980s two new nitrogenases have turned up. Both are sensitive to oxygen, both need ATP, both leak hydrogen and both comprise two proteins (the larger protein is a mite more complex in the two new ones). They are just as difficult to study and as enigmatic in their properties as is the molybdo-enzyme. One or other has since been detected in other species of nitrogen-fixing bacteria.

Today there seem to be three families of nitrogenase, all working in similar ways. Mechanisms in which nitrogen gas molecules displace hydrogen from a metal hydride might, of course, still be true of all three, with the displacement happening at iron atoms. But one can no longer be confident that metal atoms are involved at all; sulphur atoms, which are abundantly associated with iron, molybdenum and vanadium in all three kinds of nitrogenase, might bind nitrogen. For that matter, other atoms which go to make up these proteins – oxygen, nitrogen or carbon – cannot really be excluded. And again, each nitrogenase could do its thing differently.

The study of nitrogenase has made spectacular progress over

the last 30 years, with biochemistry, chemistry and genetics all contributing. Despite all obstacles, biochemists have extracted and substantially purified the enzyme proteins from a dozen or more species of nitrogen-fixing bacteria, and they have worked out the structures of molybdo-nitrogenase proteins in great detail: in 1992 X-ray analysis of the protein from an azotobacter revealed the locations of the iron and molybdenum atoms, with attendant sulphur atoms, within the large protein molecule. During the last 30 years, too, a whole new area of the chemistry of the element nitrogen has been revealed, concerned with its ability to form and interact with complex compounds of molybdenum, vanadium and iron, as well as a dozen or so related metals.

So it is ironical that, in terms of understanding how nitrogenase really works, scientists find themselves almost back where they were in the 1960s – except that they have a lot more hard facts to work on; and now they know that the term 'nitrogenase' is a collective noun, in more senses than one.

The why . . .

The complexity of nitrogenase – I shall continue to use the name, but now for a family of pairs of enzymes – does not just begin and end with the way it works; it entrains a whole worm-can of subsidiary questions.

Here is one. If most nitrogenases are molybdo-enzymes, as seems to be the case, why is there a vanadium nitrogenase at all? It transpires that the molybdenum enzyme leaks less hydrogen than the vanadium enzyme, so one could argue that it is therefore more efficient. And it is certainly true that azotobacters prefer the molybdenum enzyme: if molybdenum is available, they do not make the alternative nitrogenases. But the efficiency argument does not really stand up to close examination, because it is only true in the laboratory, where experiments are generally done at 30° Celsius. Out in the real world, in soil and water where the bacteria live, temperatures are usually around 4 to 14°, and in such cool conditions, the molybdenum nitrogenase becomes leaky. In fact, between 5 and 10°, the molybdenum nitrogenase is less efficient than the vanadium one. Could it be that, by using incubators and putting molybdenum into their culture media, microbiologists have simply ensured that they only isolate molybdenum-preferring azotobacters from the outside world? We cannot yet be sure, but if it were true, it would not be the first time that laboratory isolates have proved to be unrepresentative of the types prevalent in nature.

Here is a more general question. Nitrogenases of all kinds are extremely sensitive to oxygen, so how do bacteria manage to use them at all, surrounded as they are by some 20% of the gas? Finding the answers, for there are several, to this problem has dramatically advanced understanding of how nitrogen-fixing bacteria work.

Actually, quite a few nitrogen-fixing bacteria do not solve the problem, they evade it. This happens with a class of bacteria which grow without air (I wrote about them in Chapter 7). Some of this class can fix nitrogen, but they do not do so unless they find themselves in an environment which is essentially oxygen-free, such as in a compost heap, or a polluted pond sediment. Here the respiration of all kinds of other bacteria, as well as fungi, worms and so on, uses up oxygen as fast as it gets in and the problem of eluding it does not arise. The story of a species of bacterium called *Klebsiella pneumoniae* illustrates how our understanding that they side-step the oxygen problem advanced knowledge. This species is very easy to culture and is normally grown in air, though it can manage quite well without. During the 1950s, strains of *K. pneumoniae* were suspected of being able to fix nitrogen, but whenever scientists tried to prove this incontrovertibly, they could not. But they had been testing their cultures in air; once the oxygen problem was understood, after 1960, tests were done culturing the organism without air, and the suspect strain proved not only to fix nitrogen unequivocally, but to be among the best nitrogen fixers available – one strain of *K. pneumoniae* has since become a sort of 'work-horse' for research on the subject. All this led to the discovery of several entirely new kinds of nitrogen-fixing bacteria, ones which are able to grow without air and then, but only then, can fix nitrogen.

Conversely, there are nitrogen-fixing bacteria which take steps to get rid of oxygen themselves. Azotobacters cannot grow without air, and they were long known to be capable of enormous respiration rates: they can respire (which means consume oxygen while burning up food) 40 or 50 times faster than, for example, *Klebsiella* (and, incidentally, over 100 times faster than the cells of human beings). For many years this seemed to be metabolic freakishness – and very wasteful of food. But the property began to make sense when scientists realised that the bacteria use this tremendous respiratory activity to scavenge dissolved oxygen from around the cell, so that it cannot damage their nitrogenase. They do indeed waste a lot of food supporting that respiration, and they do not do it unless the stratagem is going to work. So they also have a fail-safe system: they possess a special protein which, if there is more oxygen around than the cell can cope with, sticks to molybdenum nitrogenase and protects its oxygen-sensitive parts. This stops the nitrogenase from functioning, but at least it saves it from disaster, and if circumstances change favourably, the protective protein comes off and the enzyme can work again.

Azotobacters must have oxygen to live, so they confront the oxygen problem directly, and solve it in the quite sophisticated manner I just described. Klebsiellas can do without oxygen, so they duck the problem. An intermediate class of bacteria comprises those which, though unable to grow without any oxygen at all, can manage with very low oxygen concentrations. In the past couple of decades, quite a few of these have been shown to be able to fix nitrogen if, for example, they find themselves in a low-oxygen environment, such as the edge of an active compost heap. But they cannot do it out where air is plentiful. Several types find a low-oxygen niche around the roots of weeds, grasses and even cereals: they are thought to benefit the plant by providing them with a little nitrogen (not much, regrettably, or they would be widely encouraged in agriculture). Perhaps the most important bacteria of this kind are those that are actually known to help the plant, the symbiotic bacteria, such as the *Rhizobium* species which co-operate with leguminous plants such as clover, lucerne, peas and beans. These have to have oxygen, and they find comfortable, low-oxygen niches, well supplied with food from the plant host, within the root nodules which they inhabit. Root nodule bacteria, especially those of legumes, are of very great importance in world agriculture and also in the natural economy of huge areas of this planet; it is a relatively new thought that those nodules are there as special compartments, to limit access of oxygen to the microbes' nitrogenase. They perform this task most ingeniously, but I must not go into such details now.

Consider, instead, the photosynthetic bacteria. These, like plants, use sunlight to make organic matter from carbon dioxide. Some do it without involving oxygen in any way, and quite a few of these can fix nitrogen. But others, such as certain organisms called Cyanobacteria, have an exceptional problem, because they not only inhabit aerated environments, but their photosynthesis, like that of plants, generates oxygen from water. Yet they also fix nitrogen. They solve their dilemma by making their own compartments: they grow as filamentous rows of cells, and they set aside special cells at intervals along the filament in which nitrogenase resides and functions. Those cells are almost impermeable to oxygen; they are called 'heterocysts' and they are very distinctive. They had been known since the late nineteenth century, but their function was wholly obscure until their role in nitrogen fixation was shown in the 1970s.

There are a few other ways in which nitrogen fixers get over their problem with oxygen, but I think that is enough for now. Before the oxygen problem was recognised there seemed to be only about half a dozen types of bacteria whose ability to fix nitrogen had been authenticated (by demonstrating the ability to fix the heavy isotope of nitrogen, which I mentioned at the beginning of this chapter). Today the list exceeds a hundred, and includes new symbioses as well as novel ways of microbial life.

A few other questions arise from the properties of nitrogenase. Is the need for ATP a serious problem to the cell? (Snap answer: not really; nitrate is only marginally less 'costly' a source of nitrogen atoms, in terms of ATP, than nitrogen gas.) How do microbes regulate the synthesis and working of this multiplicity of oxygen-sensitive enzymes which they do not want to use unless they must? (Snap answer: ingeniously, but this is an on-going research problem which it would be tedious to examine closely here.) Why does Dr Bishop's azotobacter need three different kinds of nitrogenase? (Snap answer: enough! I don't know, nor does he.) However, there is one very simple puzzle which I ought not to set aside.

The hundred-odd known types of nitrogen-fixing bacteria represent but a tiny proportion of the thousands of named types of bacteria, yet, as I wrote earlier on, they are the only living things which can fix nitrogen. Why is the distribution of the property so limited? Surely it would be a tremendous advantage to plants, for example, to fix nitrogen themselves and not to have to rely on bacteria? And would not ruminants such as goats and cattle, or wood-eating creatures such as termites, both of which live on diets low in combined nitrogen, benefit from a spot of nitrogenase in their stomachs? Why has evolution, which has thrown up creatures as complex as ourselves, who can pose the question, not generated a nitrogen-fixing plant, for example? (It has not, by the way – I mention this in case you have come across one of the bogus reports I discussed in Chapter 10.) I think I know why. I think that we are seeing just another consequence of the inertness of nitrogen and the complexity of nitrogenase.

Reflect on the early days of life on this planet, more than 3 billion years ago. There were no plants or animals; the world was warm and wet, and its waters were inhabited by primitive blobs of protoplasm, looking and behaving rather like some of today's bacteria, those which live without oxygen. For though the composition of the atmosphere in those days is disputed, authorities are almost unanimous in asserting that there was effectively no oxygen about. A little was probably formed by non-biological processes, but it would have been mopped up rapidly by geochemical reactions. Instead there was nitrogen, carbon dioxide, water vapour and small amounts of other gases. It was a turbulent world, and its microbes were adequately supplied with nitrogenous compounds formed by natural chemical means in the atmosphere, such as lightning, irradiation and vulcanism. There was no reason for living things to fix nitrogen, and the property did not arise. The sort of natural selection which would lead to the emergence of nitrogenase would begin to operate only when the Earth's inhabitants began to run out of supplies of geochemically-formed nitrogenous compounds. So when did that happen?

Nobody can be sure. Many think that it happened quite early in life's history, but my guess is that it was quite a late event: a couple of billion years after life originated, round about the time, 1½ billion years ago, when plant-type photosynthesis, which had developed among microbes resembling cyanobacteria, began to make so much oxygen that it became a permanent component of the atmosphere. At that time the Earth would have calmed down geochemically, but microbes, still the only kind of living things but by then quite sophisticated in their biochemistry, were already transforming the world into the sort of place we know today. The impact of oxygen on many of the microbes would have been catastrophic, because most would have found it poisonous and died – and much fixed nitrogen would have been recycled – but oxygen-forming creatures would have multiplied, and air-breathing, oxygen-consuming creatures would have appeared. Evolving together into ever more sophisticated creatures, they would have colonised old and new zones of the planet with their new, efficient metabolisms, displacing but not eliminating more primitive types. That early population explosion might well have caused the biological demand for fixed nitrogen to exceed supply, creating conditions for the evolutionary emergence of nitrogenase.

Why did only bacteria learn how to fix nitrogen, and not the more sophisticated precursors of plants and animals? Because, if they were anything like their representatives today, the bacteria were much more flexible than other creatures. They were much better at adjusting their genes to cope with new situations, as nowadays they prove to be when confronted with carelessly used drugs and antibiotics. (I suggest, incidentally, that it was the more primitive bacteria, the ones which did not use oxygen, which learned the trick first, and passed it on to their evolving descendants. Which is why the enzyme is still oxygen-sensitive.) Why did not more complex organisms learn the trick too? *Because for millions upon millions of years they did not have to.* That is a crucial point. Once enough bacteria became able to fix nitrogen, they would grow and die in sufficient numbers for their nitrogenous components to recycle and to supply the rest of the living world. The selection pressure necessary for more complex organisms to learn the trick disappeared again. Of course, some of those more complex organisms found it advantageous to give house-room to nitrogen fixers, and that is why we have the symbiotic systems, most of which seem to be millions rather than billions of years old. There is reasonable evidence that the few species of bacteria that fix nitrogen kept the rest of the living world supplied until the present century, when mankind's agriculture began seriously to perturb the biological nitrogen cycle.

Well, that's what I think, and you do not have to believe it. Some well-informed scientists do not – they prefer to think that nitrogen fixation is an exceedingly ancient process which, once widespread, has been lost by most organisms. But my guess – for guess it is, albeit an informed one – has a consequence that is worth contemplating. Nitrogen-fixing plants did not appear because they had no need to. But the human population explosion, with its concomitant pressure on the plant kingdom through agriculture and forestry, has changed the scene: it has produced the necessary selection pressure. Farmers ease the pressure with chemical fertiliser, but in many parts of the world such fertiliser is too expensive. The sort of selection pressure which would lead to nitrogen-fixing plants exists again, and they may now be on their way. After all, though it is a difficult process for living things to manage, all the problems presented by nitrogenase – oxygen sensitivity, ATP demand, hydrogen formation, regulation and control – have been solved by bacteria. Plants could develop comparable solutions.

That sort of thought was thought a decade or two ago. Some scientists dismissed it; others are already researching on schemes to hasten things: to transfer the genes for nitrogen fixation from bacteria to plants, in a form that the plants could use. It is now 20 years since nitrogen fixation genes from a natural fixing bacterium were manipulated into another species in the laboratory, and a wholly new type of nitrogen fixer created. Some authorities think that something of the sort has gone on naturally, among bacteria, quite often during evolution. But for many reasons, doing it in plants will be a long haul. Yet there seems to be no fundamental reason why it should not work.

Success would mean that one of life's more exotic capabilities would have been introduced in higher organisms and, like so many advances in fundamental science, it could be very useful. If something practical emerged, such as nitrogen-fixing agricultural crops, it would save fertiliser costs and ease local food shortages, a matter of tremendous importance to developing countries.

And, if I am right in my thinking, the scientists who pulled it off would simply be giving evolution a push in the direction in which it is already poised to go.

12
Getting about

Do you remember being buffeted by the wind? When it seems to come at you from all sides, chopping and changing direction incessantly? It happens especially in cities, when the wind surges and eddies round tall buildings and sharp corners; in vacant lots it sometimes creates miniature whirlwinds. Most of us have experienced such winds occasionally, but usually the wind behaves predictably: no matter how strong it may be, it comes from one direction, if sometimes in gusts. And when there is no wind, calm reigns on all sides.

Bacteria, if they enjoyed our level of sentience, would find the world very different. They live in water, even though that water may only be a film of damp – on a leaf, on a soil particle, or a person's skin, for example. As with fish, their equivalent of a wind is the flow of water, a current of some kind. But even when there is no such flow, all is not calm. The bacteria are buffeted constantly by the molecules of the water – and any other kinds of molecules dissolved in it – and they are so small that they notice it. Again, I do not mean 'notice' in the sense that we should; I mean that they react to the buffeting, albeit passively. Under the microscope, they can be seen to be constantly jogging around, making small movements in random directions, and it is impossible, in truth, to get a good look at one unless, for some reason, it sticks to a solid surface.

Tiny particles of dust in water behave in just the same manner. Their movement is called 'Brownian motion', after Robert Brown, an 'exceptionally shy but brilliant' (Charles Darwin) microscopist who first described it in 1827, causing juddering of particles within pollen grains suspended in water. It happens because, at ordinary temperatures, all molecules are in motion. Those composing solids vibrate, but those composing liquids and gases move randomly, bumping one another all the time. The hotter the liquid or gas, the more vigorously do they move and bump.

Compared with a water molecule, a bacterium is huge. Yet it is still small enough to find the random battering unsymmetrical. By unsymmetrical I mean that, over successive fractions of a second, surges of molecules will push it first in one direction, then in another, then in a third, and so on. The upshot is the buffeting I wrote of, causing the constant, random wobble one sees under the microscope.

To put all this in perspective, fish, too, are buffeted by molecules, and so are people, though the molecules which beat upon us are those of the gases composing air. But organisms such as fish and ourselves are so large that they do not notice it. Or, to be strictly accurate, they notice it only when they are hit by a tremendous and one-directional surge of molecules – which is what water flow or a wind is.

Brownian motion can be looked at as numberless tiny, random surges, and these provide many bacteria with their only way of getting about. Brownian motion does not move them far, but it jostles them sufficiently to make their immediate environment change continuously, so that the cells' supplies of nutrients and oxygen are renewed and unwanted wastes such as carbon dioxide are removed. And since Brownian movement is always supplemented by small local disturbances, currents generated by heating or cooling, or by the passage or movement of larger organisms, these bacteria find the situation adequate. They have no compelling need to move of their own volition, they bumble along successfully at the mercy of their micro-environment. Even when it seems absolutely static to us, it is turbulent enough for them.

But there are other bacteria which are less suited to such constant motion. Solid surfaces can often be better sources of nutrients than free solution, and in streams, in the guts of animals, on tree trunks and so on, it can be advantageous for a cell not to move, but to remain where it is, so that it can grab some of whatever nutrient may flow past. The problem becomes one of avoiding being washed away by currents and surges, and such bacteria become past-masters at adhering to solid surfaces. They surround themselves with coatings of sticky materials, which may be actual starchy-type gum, adhesive proteins, or combinations of both. The organisms which form plaque on our teeth are especially good examples because they are subject to regular surges of saliva. From a sugary sweet, they can knock up a coating of adhesive material with startling speed. And on a larger scale, the 'floc' of sewage purification plants comprises masses of bacteria adhering to each other, entrapping other microbes as well; in polluted water, all sorts of objects become covered in a so-called 'biofilm', a coating of adhesive bacteria which provides, in its turn, a habitat for other microbes, including algae, fungi and protozoa, to attach themselves to. Such a biofilm, in a domestic goldfish aquarium, may be initiated by greeny-blue cyanobacteria, but is soon colonised by green algae; different bacteria add brown and black flecks and the whole film is grazed by protozoa, water snails (if there are any) and even fish if it gets thick enough to be appetising. Biofilms of that kind can be the origins of many kinds of food chains.

But there are yet other bacteria which the passive life does not suit, and many types have learned ways of moving deliberately. Sometimes, on a suitably slippery surface, whole clusters of bacteria seem able to swarm as one, forming a moving raft of microbes. In a freshwater biofilm, under the microscope, one may see bacteria which glide over and through the film like slugs, or others which writhe their way through it, like snakes, and one always sees bacteria which can swim. They dart in and out of the scene, sometimes tumbling about, sometimes stopping and wobbling a while, only to take off again. Microbiologists call bacteria which can swim 'motile', not the more familiar 'mobile'. (I have never been clear why; both words have the same Latin root.) A specimen of mildly polluted water, under the microscope, presents a scene like a swarm of mosquitoes dancing in the setting summer sun.

Is their bewildering movement purposeful? Probably it is. For motile bacteria seem able to congregate in environments which are especially favourable as far as nutrients, air, acidity and so on are concerned. How they know which way to go is an interesting question which I must set aside just now; my present concern is, how do they swim at all?

Under an ordinary microscope, most motile bacteria look no different from non-motile ones: they have no obvious organs of locomotion. But as long ago as 1838 the German biologist, C. G. Ehrenberg, watching an especially large and beautiful purple bacterium called *Chromatium okeni*, saw that it had a long, curly tail at one end of its oblong body, which moved rapidly and acted as a sort of propeller. That organism can race forward, stop abruptly, reverse or change direction suddenly, as another German scientist, T. W. Engelmann, described in 1883. Engelmann was interested because *Chromatium* can photosynthesise (which means convert carbon dioxide to organic matter rather as plants do, though the way it does it is very different from plants in detail), and to this end it needs light; its movements convey the cell towards light. Microbiologists guessed that many other motile bacteria possessed comparable tails, too thin to be seen, and soon specialised ways of staining dead bacteria revealed that this was true. Decades later, the electron microscope was invented and the tails could be seen and studied more clearly.

They called the tails 'flagella' (singular: 'flagellum') from the Latin for 'whip'. They were generally very long, five to ten times the length of a cell, and wavy rather than straight – but since they were seen clearly only on dead cells, microbiologists could not be sure how wavy they were in life. If the cell was rod-shaped, the flagellum would often be at one end, or there might be a cluster. Ehrenberg's *Chromatium* had a clutch of 40 flagella at one end, a bundle thick enough to enable him to see its 'tail' under an ordinary microscope. Some bacteria had flagella at both ends: some had them all round the cell, perhaps few, sometimes many. Different species had their characteristic arrangements of flagella, and spherical, bean-shaped and spiral bacteria displayed a comparable variety of arrangements of flagella, depending on the species. Bacteria with flagella, or a single flagellum, at one end were able to move rapidly in straight lines, and could reverse abruptly; those with flagella all over tended to get around with a stumbling, tumbling motion.

The existence of flagella, invisible under the ordinary microscope but demonstrably present, provided a partial answer to the problem of bacterial motility. But how, precisely, could a very long tail permit such rapid stop-start and reverse motions?

Protozoa seemed for many years to provide the answer. They are single-celled microbes, but very different from bacteria. They are much larger; they have a complex anatomy which can be seen clearly under a decent microscope, and many have flagella – thicker and more structured than those of bacteria. In the late nineteenth century microscopists, among whom Otto Büschli of Germany was notable (the Germans made excellent microscopes!), discovered that the flagella of many motile protozoa would assume a spring-like form and, by contracting and expanding, would produce a wave-like pulse which would move the cell along, pushing or pulling according to whether the pitch of the spring was right or left-handed. As an underlying principle, this was correct, and today the biochemistry of how the material composing a protozoal flagellum contracts and expands is fairly well understood.

For many decades microbiologists considered it logical to assume that bacterial flagella also worked in a comparable manner: that they had a spring-like form and transmitted a wave. But there were dissenters. I recall a distinguished South African microscopist arguing strongly, in the 1950s, that flagella contributed nothing to bacterial motility; that they were streams of mucilaginous material, from the cell's outer coating, trailing as a result of the microbe's motion. In the state of knowledge at that time, this view was just about tenable, but evidence to the contrary mounted and by the mid 1970s no-one could seriously doubt that flagella were organs of locomotion. However, the evidence did not always support the widespread assumption that bacterial flagella functioned like those of protozoa. Gradually microbiologists, especially H. C. Berg of the University of Colorado, USA, came to realise that the bacterial flagellum is a remarkable piece of biological machinery, exploiting a principle which has vanished from higher organisms – until it was re-invented by mankind as a real machine.

A typical bacterial flagellum, we now know, is a long, tubular filament of protein. It is indeed loosely coiled, like a pulled-out, left-handed spring, or perhaps a corkscrew, and it terminates, close to the cell wall, as a thickened, flexible zone, called a hook because it is usually bent. The flagellum is kept fairly rigid by its tubular structure, but the hook is more flexible, so the external set-up resembles a stiff wire corkscrew which can wave around because it is set in a flexible sleeve. One can imagine a bacterial cell as having a tough outer envelope within which is a softer, more flexible one, and inside that the jelly-like protoplasm resides. The flagellum and its hook are attached to the cell just at, or just inside, these skins, and the remarkable feature is the way in which they are anchored. In a bacterium called *Bacillus subtilis*, which has a fairly simple structure, the hook extends, as a rod, through the outer wall, and at the end of the rod, separated by its last few nanometres, are two discs. There is one at the very end which seems to be set in the inner membrane, the one which covers the cell's protoplasm, and the near-terminal disc is set just inside the cell wall. In effect, the long flagellum seems to be held in place by its hook, with two discs acting as a double bolt, or perhaps a bolt and washer.

Most of the detail I have given was worked out a dozen or more years ago using electron microscopes. It immediately raises the question, how does a stiff corkscrew-like appendage provide reversible motion? In principle, there are two possible answers: either it is waved about, and generates thrust or pull because of its spring-like shape; or it spins, functioning like a peculiar kind of propeller.

Rotating propellers are not the ordinary stuff of biology. But despite its implausibility, the second possibility seems to be the right one. When the flagellum is working, it behaves just as if the basal discs were wheels on an axle: as if they (or perhaps just one of them) spin round and, transmitting torque to the rod, spin both it and the rest of the flagellum. This, with its corkscrew-like shape, generates thrust which propels the cell forward. And whenever some stimulus or obstacle urges the cell to reverse, it at once reverses the direction of rotation, so that flagellar thrust is converted to flagellar pull. The cell itself actually rotates too, but only slowly; being much larger it is much more restrained by the viscous drag of water.

Evidence, you ask? Well, it is true that no-one has actually seen this happen while a bacterium swims freely, and it is difficult to imagine how anyone could, but the next best thing has been observed, as I shall shortly tell.

Much of the definitive work has been done with *Escherichia coli*, a species of motile bacteria which normally resides in mammalian guts. Its flagellar arrangement is not the most simple: it has several, distributed unevenly around its short, rod-like cell. These cause it to take a jagged, seeking path when swimming freely: darting forward, stopping abruptly and changing direction repeatedly. This happens because its various flagella change direction in an uncoordinated way. But it is also capable of uninterrupted, seemingly deliberate, progressive motion, and when this happens, all the flagella spin in the same direction in unison, the hooks bending so that the flagella are approximately aligned – this can happen without their getting entangled because all the coils are of similar pitch. *E. coli*'s flagellar structure is also not the simplest, in that its hook terminates in a rod with four rather than two discs. But it has made sense to study *E. coli* rather than a simpler organism because a very great deal is known about its general biochemistry, physiology and genetics, and this has provided a basis for working out how motility fits into the rest of the organism's metabolism. *Salmonella*, a close relative of *E. coli*, is very similar and has been much studied, too.

Convincing evidence for rotation of the flagella was obtained over a dozen years ago because it became possible to cause minute plastic beads to adhere to flagella, and then actually to see the beads go round and round, as linear arrays, under a conventional microscope. Even more impressive, it was also possible to make *E. coli*'s flagella sticky, so that they adhere to a solid surface and thus tether the cell; when this was done, the whole cell could be seen to rotate – just as would happen if, to use Professor Berg's simile, you fixed an electric fan to the wall by its blades and switched it on: the whole drive would then rotate. (Don't try it!)

It now seems that most bacterial flagella work in this way. A single flagellum will provide simple forward, pause and reverse movements; bundles of flagella rotating in unison will work similarly; and scattered flagella, by changing their direction of rotation at random intervals as in *E. coli*, will cause the tumbling, stumbling movements which microscopists see among many kinds of bacteria. And when there is a directional stimulus, such as if a nutrient is sensed, scattered flagella will rotate more co-ordinately and the cell will move directionally. But how do they manage to rotate?

Ships with propellers have an engine aboard. This must be firmly fixed, and its shaft must rotate freely, extending outside to the propeller through a fixed gland, which must also keep the sea water out. On a minuscule scale, motile bacteria must solve similar problems. In *Salmonella*, and probably in *E. coli* too, the two basal discs closest to the hook are embedded in the cell wall and act as a gland, which the rod penetrates and within which it rotates. The terminal disc is tightly fused to the end of the rod and is located in the protoplasmic membrane; and it is free to rotate, probably together with the fourth disc just above it. All this has been deduced from sophisticated electron microscopy of cells and of carefully prepared fragments of cells. In addition, mutant bacteria defective in parts of the flagellar apparatus have been discovered, some of which cannot move, or move in aberrant ways, and, from a study of these, many of the 40 or so genes which tell an *E. coli* how to grow the flagellar apparatus have been identified.

If the rod is the rotor of the flagellar motor – motor is the correct technical term for the biological set-up which makes the flagellum spin, incidentally – then the terminal disc must be part of the drive. And so it seems to be. Motors must have a power source, for they consume energy, but nothing equivalent to cylinders and pistons exists in bacterial cells to provide the impulse. Protozoa (I remind you) use wave motion within their flagella to generate movement, and to power the waves they make use of biological energy in a conventional way: their flagella include enzymes which actually utilise ATP (ATP is the fuel which powers the internal workings of living cells). In contrast, the flagella of bacteria do not consume ATP. They continue to function when biochemists apply certain technical tricks which interrupt their supply of ATP. How, then, does their little motor keep spinning? No-one can be sure, yet, but there is good evidence that it works as a kind of an electric motor. Differences in electric charge exist between the underside of the terminal disc and the surface of membrane beneath it; these differences generate a current flow which causes the disc to spin. On the membrane, the sites of potential difference are probably localised as up to 16 generator sites, arranged as a ring, a system which might well generate reversible torque and spin the rest of the flagellar structure. The source of the necessary electrical energy, at the molecular level, is the electrical potentials which have long been known to exist across the membranes of living cells and which are involved in the movements of charged atoms and molecules – known as ions – in and out of cells.

Flagellate bacteria, then, are powered by miniature electric motors coupled to corkscrew-like impellers: highly manoeuvrable devices. The basal discs and their rod are a marvellous piece of propellant micro-machinery, functioning on a molecular scale, and it works in water. What is especially peculiar about it, as far as evolution is concerned, is that the terminal disc is a wheel, one which rotates. Yet wheels do not occur among more highly organised living things. It is fair to say that most bacterial properties, both metabolic and structural, turn up, even if in re-invented forms, in higher organisms, but wheels have not appeared again in evolution. Yet the ability to rotate an organ through a true circle would surely have been useful, especially if it had led to the use of pairs of wheels. One wonders what sorts of living things would inhabit this planet today had natural selection elicited a creature in which homologues of the basal discs became an organ of locomotion; one in which the rod between the two pairs of basal discs extended to form a true axle, and the flagellum as such was dispensed with. The power source could have been scaled up, as the existence of the electric eel tells us. Evolution invented all sorts of sophisticated organs of locomotion: flippers, fins, legs with articulated joints, wings, devices for swinging, squirming, slithering, wriggling and jet-propelling; not to mention passive means of dispersal such as the hooks, parachutes and catapults of seeds. But it forgot about the wheel.

It has been said that civilisation dawned when mankind invented the wheel. Not quite right. We re-invented it. And look at our streets today. Perhaps Nature was right, after all, to forget.

13

Microsenses

'Hering, sight, smelling and fele, cheüing er wittes five.' Thus wrote the anonymous author of *Cursor Mundi*, a middle-English poem, in about 1300. By the mid seventeenth century the 'five wits' had become the five senses, and the belief that there are five has become so well established that today everyone takes the phrase 'a sixth sense' to mean a perception, real or imagined, which is not quite natural. Yet the rule of five has almost no scientific basis. True, the five senses are matched, in higher animals such as ourselves, by appropriate anatomical features – ears, eyes, nose, skin and mouth – but otherwise they do not stand up to even superficial analysis. Taste and smell are inextricably mingled, as even an unsophisticated wine bibber will agree; touch encompasses sensations of pain, heat and cold as well as feel; hearing verges upon a sense of pressure. And where does our sense of balance fit in? Well, perhaps it is just a special combination of sight and touch. So what about our sense of direction? Pathetic it may often be, especially in the dark, but is it not real? Sight and memory, you say? Well, I shall not argue. Try something difficult: what about our sense of time? Unreliable it may be, and we all learn that it slows down as we grow older (so time itself seems to accelerate), but is it not a real sense for all that? The message is that analysing our modes of perception is a complex task even for specialists; there is no need to stop at five, seven, or even at twenty senses.

Biologists have to think about senses, because all living things need to perceive their environment in one way or another. For example, the sorts of animals which move around, feed, fight (or evade), mate and so on, have to be pretty good at perceiving. Sometimes, like the blind mole, their habitat is so special that they can dispense with the odd sense. Some have special senses, such as fish which must sense hydrostatic pressure to control their depth. Some creatures develop familiar senses to an astonishing extent: the dog's sense of smell can be impressive, but it is nothing compared to that of a male moth scenting a female from a couple of miles away. The rabbit's hearing is famous, but bats, with their high-pitched sonar, have extended hearing into a form of night vision. And the sense of direction of the pigeon is legendary. At the other extreme, even a plant, which many would regard as insensate, has to 'know' which way is up, so that its stem grows upwards and its roots downwards. Most plants can sense, and respond to, light, or drought; and all 'know', with considerable precision, what nutrients to extract from the soil.

Bacteria are no different from other living things in having to react to their environment, so they must also perceive it. But how can we be sure they do? And if we can be sure, how much of their surroundings can they sense? And how do they do it?

They live in water, even if it is only an imperceptible film of water on skin, on a leaf, or on a soil particle, and it is the behaviour of bacteria which are able to swim about that has answered, or partly answered, these questions. I told in Chapter 12 how bacteria such as *Escherichia coli* swim. To save you looking back, I shall summarise: their cells have corkscrew-shaped propellors, called flagella, powered by what amount to tiny induction motors, which spin around to propel the cell forward, and reverse spin to stop and go backwards. Each cell has several flagella, and these may function co-ordinately to give progressive motion, or out of phase to give a seeking, tumbling motion. Bacteria exist which have only one flagellum, and others which have tufts of them, usually at one end; but the flagella all operate in similar ways – as far as scientists know – and all can swim about. They are 'motile', in technical jargon.

The sort of thing microbiologists do to study bacterial senses is to get a population of motile bacteria swimming peacefully and randomly in a drop of watery medium, perhaps under a microscope, and then introduce some change at one place – with a capillary tube perhaps, containing something nutrient or nasty. Then they watch, to see if the bacteria swim away from the disturbing substance, towards it, or pay no attention. There are several variations of this general approach; sometimes it is advantageous to have the bacteria in capillaries. Sometimes their movements can be filmed or monitored electronically. A microscope is not always needed: sometimes bacteria will visibly cluster in a favourable environment, as I shall tell later. By such methods, the ways in which motile bacteria respond to changes in their environments have been studied for over a century, with ever increasing sophistication.

The variety of sensory abilities bacteria can display is impressive, but I must make it clear straight away that they do not – they cannot – integrate their various sensations in the way in which we, and even quite primitive higher organisms, can. Nor does one find all the bacterial senses in one species. The movements bacteria make in response to a stimulus, towards or away from the source as the case may be, are called 'taxis'. The word is pronounced *tacksiss* (not as if it were the plural of a taxi), and it is a term I shall have to use quite often. It usually has a prefix indicating the cause of the movement: phototaxis, for example, means movement provoked by light.

Probably the best understood kind of taxis is chemotaxis: an ability to sense and respond to chemicals. These may be either beneficial ones, nutrients in particular, or detrimental ones such as toxic substances. T. W. Engelmann, a German microscopist who was a pioneer in this area of research, first described chemotaxis in 1881. His description, and that of another German contributor, W. Pfeffer, in 1883, established an important feature of chemotaxis: that an organism such as *E. coli* can show either positive or negative chemotaxis. This means that it is attracted by some substances and repelled by others. In 1979 Julius Adler and his colleagues of the University of Wisconsin at Madison, USA, published a fascinating survey of the likes and dislikes of *E. coli*: most nutrients, such as sugars and amino-acids, are attractants, and many mildly toxic substances are repellents. In addition, acidity and alkalinity are repellent, so the organism always tries to home in on a neutral environment. *E. coli* cannot digest one of the most common sugars, sucrose, and this proves to be neither an attractant nor a repellent: *E. coli* does not respond to it. Clearly there is an elementary logic to *E. coli*'s predilections, but the machinery of its responses is not quite so simple.

I have so far implied that *E. coli* swims towards an attractant. Actually, that is misleading. What *E. coli* really does is to swim in a random manner, but, if it finds itself swimming away from where an attractant is plentiful, it pauses, and then changes direction. Because *E. coli* cannot reverse, but does a right-angle shift instead, the path it follows responding to an attractant is jagged. But it is led to the source of its attractant very effectively. Sometimes one sees uninterrupted linear runs, but they are quite rare. Spirilla, which are corkscrew- or S-shaped bacteria with their flagella at one end of the cell, travel less jagged paths because they can reverse precisely. But the bacterial engine rooms, metaphorically speaking, are not instructed to go ahead, they are only instructed not to go ahead. The cell goes away from a *lack* of attractant, just as it goes away from a repellent. Chemotaxis is basically a negative response to unfavourable conditions, even if it leads to results which seem to be positive.

Research is only too often a long and tedious slog, even if its results can be surprising and exciting. In recent decades combined research by many laboratories has enabled scientists to work out in remarkable detail how *E. coli* senses a substance and reacts appropriately. An *E. coli* cell is like a stubby little sausage, but one with two skins. The inner skin is rather fragile and encases the cytoplasm, and the outer one is tougher: a protective coat which also maintains the cell's sausage-like shape. Between the two skins is a narrow space in which some important parts of the microbe's metabolic machinery reside. Average-sized molecules, by which I mean sugars, amino-acids, salts and the like, but not big molecules such as starch or proteins, can penetrate the outer cover and enter this space. If a molecule gets in which is one of *E. coli*'s attractants, it is detected by a special protein molecule awaiting it, which is embedded in the inner skin and is large enough to reach right through it. It is called a 'receptor' or 'transducer' protein. (Some kinds of molecule first combine with a protein waiting in the space between the skins which conveys them to the appropriate receptor, as I shall call it.) The receptor has a socket in its molecular structure which just fits the molecule being sensed – like a key fitting a lock – and the molecule sticks to the protein at that site. This event causes the other end of the receptor protein, the end in contact with the cytoplasm, to change its shape slightly, provoking quite a cascade of reactions. An enzyme waiting in the cytoplasm modifies the exposed part of the receptor protein, by attaching little molecular fragments called methyl groups at certain points; two other proteins react with each other in turn, phosphate groups coming off one protein and attaching to the next, and finally, probably by attachment of one of the latter proteins, the flagellar motor is instructed to keep going.

There is also a loop of reactions whereby the original receptor protein is relieved of the molecule it has sensed, and (rather more slowly) the methyl groups are removed: the sensor is re-set. This loop is very important. It 'clears the decks' for the next molecule. But it also does something one would not expect in so primitive a creature: since not all the methyl groups are necessarily removed before another attractant molecule is sensed, their number provides an element of short-term memory, which is needed when a change of direction is called for. For example, the cell would have to realise that an attractant which it perceived a moment earlier is no longer so plentiful, and make a change of course. Even bacteria, like people and computers, have to have memories of a kind.

There is also machinery enabling the cell to get used to a

constant level of attractant, and a more or less converse machinery for sensing repellent molecules. All this has been worked out by a combination of biochemistry and genetics – there are over 40 genes carrying the information needed for sensing and motility in *E. coli*, at least ten of them concerned with chemotaxis, and research has been greatly helped by the discovery of genetic mutants which are defective in various steps of the sensory and response pathways. *E. coli* has no special sensory area or site on its inner surface: nothing corresponding to a nose or tongue exists. Receptors appear to be distributed all over the inner membrane, and they feed their information into the cytoplasmic cascade system from all over. Some receptors are very specific, sensing only one kind of molecule; others sense classes of molecules. But the upshot of the whole process is that, if there is something nice wafting towards *E. coli* in the surrounding fluid, the microbe will manipulate its flagellar motors so as to progress seekingly to meet it; if there is something nasty abroad, it will reverse thrust, tumble and set off in a new direction, ultimately achieving a jagged path away from the noxious substance.

E. coli belongs to a large group of bacteria which breathe oxygen but can do without it if necessary. However it much prefers to have oxygen, and it is able to sense it: it displays what for obvious reasons is called aerotaxis. Again it was Engelmann who first described aerotaxis, though not in *E. coli*. Perhaps his most elegant experiment was to set up his bacteria on a microscope slide, in water, together with some live green plant cells. When the slide was illuminated, the plant cells produced a little oxygen, from photosynthesis, and he could see the aerotactic bacteria cluster around them. Modern work has disclosed that aerotaxis is not quite the same as chemotaxis: oxygen molecules are generally sensed differently. This is known because certain, though not all, mutants defective in chemotaxis still show aerotaxis. Oxygen is needed for respiration, and it seems to be that feedback from the respiration process, rather than from an oxygen receptor, sends a signal to the flagellar motors. An important part of bacterial respiration leads to a flow of electrically charged particles across the inner membrane, and in *Salmonella*, a close relative of *E. coli*, an aspect of this flow, the flow of positively charged particles (protons), has been shown to be linked to part of the chemotaxis cascade. Too much oxygen is damaging to both *E. coli* and *Salmonella*, and oxygen can become a repellent at high concentrations; when this happens it is sensed by a different, unknown, mechanism. Spirilla belong to a class of bacteria which prefer low oxygen concentrations; in their case the attractant and repellent effects of oxygen are finely balanced and, in an undisturbed culture in a tube under air, they will all migrate to the same level, forming a visible zone some way down from the meniscus. Anaerobes, which are bacteria which cannot use oxygen and find it toxic, treat oxygen as a repellent and move away from it: they show exclusively negative aerotaxis.

Photosynthetic bacteria are creatures which, like plants, use light to make carbon dioxide into carbohydrate, and so provide their own major nutrient. So they show the phototaxis which I mentioned earlier: in general, they seek illuminated places. They are red or purple as often as green but, as in the leaves of plants, this is just a matter of the balance of pigments; their photosynthesis still depends on the green pigment called chlorophyll. Some, the cyanobacteria, are just like plants in that they form oxygen when they photosynthesise, but the ones which have been most studied from the point of view of taxis have a more primitive photosynthetic equipment, one which does not make oxygen and, indeed, does not work if oxygen is present. In the dark, these bacteria can grow with air, but they need organic food to do so. (Incidentally, they then show aerotaxis just like *E. coli*, but in light aerotaxis is suppressed.)

As in chemotaxis, phototaxis seems to be a negative thing: photosynthetic bacteria swim away from dark rather than towards light. If a tiny circle of light is shone through a culture, the bacteria congregate in it, but under the microscope they can be seen to be in constant movement, often reaching the edge of the illuminated zone, stopping suddenly, and starting off in another direction. Engelmann, who first reported this behaviour, called it 'shock movement', a name still used because, watching them, one could imagine that they are trapped in the zone of light and, seeking to escape, are shocked into giving up when they reach an invisible barrier at its edge.

In a classic experiment in 1919 a German botanist, J. Buder, showed that, if the illumination of a culture is split with a prism into a spectrum, purple bacteria cluster in three zones in the red area and, less enthusiastically, in three or four zones where green merges into blue. The three zones of red light match the three wavelengths which are used by their chlorophyll (chlorophyll is green because it absorbs red light). The blue-green zones correspond to wavelengths of light absorbed by the pigments – carotenes – which make the cells purple; in the cell these help photosynthesis because they pass the energy of the light they absorb to chlorophyll. You might deduce that chlorophyll, carotenes, and/or the rest of the photosynthetic machinery, are actually involved in sensing light, but this is not so. In 1976, by growing such bacteria with air in the dark, S. Harayama and T. Iino, of the University of Tokyo, obtained cells that were free of chlorophyll and the rest, yet they still swam towards light. The nature of the light sensor in these bacteria is still not clear.

One interesting group of bacteria, called the halobacteria, grow in very salty water. Though I wrote about them in Chapter 5, I shall remind you of the point that is relevant just now: they can make use of light by way of pigments related to those found in the eyes of higher animals. They do not make carbohydrate from carbon dioxide as plants and photosynthetic bacteria do, but they use light as an energy source. They, too, can sense light, and in their case their preferred wavelength is green. Blue light, which damages their light-absorbing pigment, repels them. But again it does not follow that their light-absorbing system includes the actual sensor.

Curiously, even our old friends *E. coli* and *Salmonella*, though they make no use of light at all, are sensitive to high illumination, and they tumble more frequently and make avoiding responses in strong light. There is a certain logic in this, because strong light damages certain of their ordinary cell components, especially molecules called flavins. Perhaps these act as sensors? No-one is sure.

Bacteria can also sense temperature: they show thermotaxis, as the pioneer Engelmann reported in 1883. If you were to observe populations of *E. coli* swimming around with no gradients of attractants and repellents but at different temperatures, you would notice the bacteria moving faster the warmer they were. Cool a culture slightly but suddenly, and they all start seeking and tumbling about, seeking warmth, no doubt. Warm them again, and they quieten down: seeking diminishes markedly, If you were to take a tubular culture of *E. coli*, on its side, and warm one end to 40° Celsius, and keep the other cool at 10°, so that there was a temperature gradient along the tube, you could see such differences in velocity and seeking pattern along the tube; and in time the bacteria would all cluster at about 34°. How do they 'feel' the right temperature? Recent work by Y. Imae and his colleagues at the University of Nagoya in Japan indicates that one of *E. coli's* chemotaxis receptors, the sensor for the amino-acid serine, can double as a thermometer: warmth brings about the same sorts of changes, including addition of methyl groups, as does sensing serine. Among the lines of evidence are that too much serine blocks thermotaxis, and that chemotaxis-defective mutants have defective thermotaxic responses. Thermotaxis, then, is closely linked to the chemotaxis cascade.

Bacteria have another, rather specialised sense, which informs them about how watery their environment is. This may seem odd, given that bacteria live in water, which is unequivocally wet by human standards, but the point is that, to creatures that live in watery places, the water is actually diluted perceptibly by other molecules dissolved in it. If the water is quite salty, as in the sea, the salt actually dilutes the water so much that it shrivels up sensitive bacteria: by sucking water from their protoplasm it dehydrates and kills them. (That is why salt is a preservative; strong sugar solutions can do the same thing, as in jam.) Conversely, very pure water does not suit bacteria because the substances dissolved in their protoplasm tend to suck in water and swell up the cells too much. Scientists measure the dehydrating effect of dissolved material in water by a property called the 'osmotic pressure' of the solution, which provides the word 'osmotaxis', meaning bacterial movement in response to the physical effects of dissolved matter. The fact that bacteria retreat from too much or from too little dissolved matter was reported by a Belgian scientist, J. Massart, in 1889 and 1891; in the 1980s Adler exploited his 'shopping list' of *E. coli* to re-investigate the matter. Having shown that sucrose, a sugar, and ribitol, a sugar-like substance, were neither attractants nor repellents for *E. coli*, he exposed these bacteria to sources of these substances at various concentrations. It transpired that they sought out an environment of a certain, moderate, osmotic pressure which was the same with both substances; mutants defective in the chemotaxis cascade which I described earlier nevertheless showed osmotaxis, implying that not only the 'osmosensor' but also its means of communicating with the flagellar motors are different from the chemotaxis system. What and where the osmosensor is, and how it acts, are still obscure.

There has been a logic to the bacterial senses that I have discussed so far: they enable the organism to recognise and seek nutrients, or favourable environments. and to avoid unfavourable substances or places. Here is one with a less clear utility. As recently as 1975, R. P. Blakemore of the Woods Hole Oceanographic Institute, USA, discovered a class of bacteria which exhibits what is called magnetotaxis: the bacteria sense the Earth's magnetic field. They align themselves with the field and move, northwards in the northern hemisphere, southwards in the southern hemisphere. (And, I fondly believe though I have never seen it, wildly in all directions at the equator.) They comprise about half a dozen species and are rather easy to isolate because an ordinary magnet will override the Earth's field. If you place the correct pole of a magnet in a culture, or even in mud or a pond which contains them, they migrate towards the pole and cluster round it. They can even influence each other magnetically: dense populations of magnetotactic bacteria tend to line up alongside each other and move in bands. They behave as if they carried tiny little compasses inside themselves.

And so they do. When they die, they remain magnetic (indeed, much of the magnetism of sedimentary rocks is apparently due to the lined-up residues of magnetotactic bacteria), and it transpires that, within their cells, magnetotactic bacteria carry tiny, very uniform, crystals of magnetic iron oxide (magnetite, the stuff of the mineral lodestone, which was the earliest kind of magnet). One species carries magnetic iron sulphide. These tiny magnets align themselves in a magnetic field and, in some way which is at present quite obscure, the microbe adjusts its direction of movement to the alignment of its compasses.

What possible use can this ability be to bacteria? It is a difficult question, and some microbiologists believe that it is useless: that some accident of their metabolism causes them to deposit iron oxide, which just happens to be magnetic, in their protoplasm. Others have proposed rather implausible ways in which iron oxide, magnetic or not, could be useful. Blakemore's own suggestion arises from the fact that magnetotactic bacteria all dislike air. Some are anaerobes, and the rest are what are called micro-aerophiles: bacteria that prefer much less oxygen than is present in well-aerated water. Blakemore realised that, though we humans regard the magnetic lines of force as coming from the north and south, to the bacteria they come out of the ground. So to follow the lines towards their source would lead the bacteria along a direct route away from air, down into the mud or sediment. It is a plausible idea, which accounts for the different directions of bacterial migration between the northern and southern hemispheres. But how do they know when to stop? A. M. Spormann and R. S. Wolfe, working at Marburg, Germany, concluded in 1984 that they possess aerotaxis sensors able to override magnetotaxis when they reach the right place. Magnetotaxis, then, seems to be a useful sense of direction, except, as I indicated earlier, at the equator, where the lines of force are horizontal.

Adler, whose researches have revealed so much about *E. coli*'s sensitivities, has looked into yet another bacterial sense, one which was described as long ago as 1889 by M. Verworn of Germany. It causes galvanotaxis which, as you will guess from the name, is movement in response to an electric field. If two electrodes are inserted into a culture of *E. coli* and a current, weak enough to avoid heating, is passed, the bacteria swim towards the positive pole, the anode. If the polarity is reversed, they do what amounts to a U-turn and seek the anode again. Dr Adler also has a strain of *Salmonella* which seeks the cathode. Tests with mutants defective in the chemotaxis genes showed that the receptors for chemotaxis are not involved. As with magnetotaxis, a

feature of galvanotaxis persists after death: bacteria killed by brief
boiling align themselves in an electric field, and flip if the polarity
is reversed. A number of animals, ranging from protozoa to fish,
show galvanotaxis, yet an ability to sense an electric field seems
wholly useless – except perhaps among creatures at risk of
encountering electric eels. There are many physiological processes which involve small differences in electrical potential
across *E. coli*'s two membranes (I mentioned the example of pro-
ton flow earlier). It could well happen that the upshot of all these
potentials is a net polarity between one end of the cell and the
other. This would cause the bacteria to align themselves passively
in an electric field and would constrain them, if they move, to
move along it; the direction, but not necessarily the strength, of
this dipole might well be retained when the cell is boiled, so that
it would still align itself. As far as bacteria are concerned, I am
inclined to suspect that galvanotaxis is a fortuitous property, with
neither a sensor nor any biological function: a taxis with no corresponding sense.

Finally, in contrast, here is a sense for which there is no evidence of taxis. Many types of bacteria are 'tactophilic', which
means that they seek out and stick themselves to a solid surface.
Other types can be seen to reverse direction when they encounter
an obstacle. Do they have a sense of touch, so that they 'know'
when they have hit an appropriate surface? Or could it be that
they hit at random, to stick or bounce as the case may be? Surface
adhesion and avoidance play a very important part in the lives of
many types of bacteria, but how they seek and recognise a solid
surface is a question that has not, to my knowledge, been studied
in detail.

Well, galvanotaxis, and surface-seeking, may be fortuitous, but
I still find it astonishing that creatures so tiny and lacking in
organ-like structures can be aware, in some way, of such a variety
of details of their environment. Within the cubic micrometre or
so of living matter that constitutes *E. coli*, a million millionth of
the volume of a small sugar lump, are primitive versions of several
of the senses that one finds in higher organisms: chemotaxis
(repesenting taste), aerotaxis (representing smell), thermotaxis
(representing warmth and cold), phototaxis (representing sight),

magnetotaxis (representing direction), possibly even galvanotaxis. And who yet knows whether sound (representing hearing) and solid surfaces (representing touch) are not sensed? Life has more diverse senses to offer than we humans may think, even if they are more mundane than the extra-sensory perceptions so beloved of mystics.

Perhaps it is just as well that we lack some, for there is enough sensory turmoil in our everyday world as things are. Just think of having to be aware of the electromagnetic fields, from TV, radio, the power grid, sparking plugs and so on, that surround and permeate us all the time.

14

A private space

Then

Let me start right at the beginning. Once upon a time there was no life on Earth. Our planet came into being, condensing out of some cosmic rubble that surrounded the sun, about 4½ billion years ago. Scientists have worked out that age from the rate at which certain unstable chemical elements, the radioactive ones, break down. It is reasonably reliable as such numbers go – which means give or take a hundred million years or so. Scientists also know that the very young Earth was too hot and dry for even the most hardy of terrestrial life forms to exist. But by 3½ billion years ago it had cooled down and life had appeared – if we accept the evidence of certain very ancient rocks, which bear traces of what look like fossil bacteria. During the intervening billion years, then, the planet became inhabitable and inhabitants duly appeared. How?

These are questions which have preoccupied philosophers and theologians since mankind started to think, but only in the mid twentieth century did it become possible to propose answers to either question which were scientific, rather than mythological or purely speculative. I shall set aside the question of how long the planet took to become inhabitable, except to state the obvious: it had to cool to below the boiling point of water, because life is wet. As far as how life originated is concerned, the basic experiments were done early in the 1950s by Harold Urey and his collaborator Stanley Miller of the University of Chicago: they showed that all sorts of organic compounds were formed when a wet gas mixture resembling the Earth's most primitive atmosphere was exposed to electrical discharges or ultraviolet light. They analysed the products and discovered that the majority were substances of the kind that are the building blocks of living matter today. For example, carbohydrates, including the cellulose of plants, are made up of sugar molecules strung together. Proteins, which provide most of life's machinery and much of its structure, are bundled-up chains of molecular units called amino-acids. Nucleic acids (RNA and DNA – I shall explain about these shortly), which hold and transcribe the genetic information which determines what a cell will be and do, are also chains of molecular units strung together, slightly more complicated than the others because they are built up of three types of unit: sugars, molecules called bases, and an inorganic component. The salient fact, in its time remarkable, is that almost all of the sugars, amino-acids and bases which today constitute living things were present in the mixtures generated by Miller, Urey and their successors. They deduced that organic matter of that kind would have been formed almost continuously on the primitive Earth because there was no ozone layer to screen out ultraviolet light and, moreover, lightning would have been an almost constant feature of the hot, wet and turbulent climate: a highly effective electric discharge. They also proposed that the molecular building blocks of which living matter now consists were the chemicals most readily available at the time of life's origin.

Actually, the gas mixture that Miller and Urey chose to use was logical enough at that time, before space probes had changed our views of planetary atmospheres, but it is no longer regarded as representative of the Earth's pristine atmosphere. However, not only have their findings been abundantly confirmed in many laboratories throughout the world, but their methods have been shown to work using all the sorts of gas mixtures which can plausibly be proposed for that early period. Moreover, though electric discharges or ultraviolet light generate such compounds well, they are not essential: local heating, such as might occur around a volcanic crater, is also effective. The one circumstance which seems to put a stop to this process is the presence of gaseous oxygen, but scientists are unanimous that free oxygen did not appear in significant amounts on this planet until after life became established – most scientists would say not until a couple of billion years afterwards.

Experiments of the kind that Urey and Miller performed tell us that the building blocks of terrestrial life arose by spontaneous chemical processes as soon as the planet cooled sufficiently for liquid water to be reasonably abundant. But they do not tell us how those building blocks came together, and became sufficiently organised to form autonomous units and, more critically important, to form units which could reproduce themselves. That question is still open, but over the last few decades information bearing on it has gradually accumulated: it is now possible to propose logical scenarios for the spontaneous origin of terrestrial life in the sort of conditions that probably existed some 4 billion years ago.

The first problem one has to face is a simple matter of dilution. The compounds observed by Miller, Urey and their successors (henceforth I shall call them Miller–Urey substances) are formed in small amounts. Even under a régime of constant electric storms, rain would wash them into seas and lakes, where they would become so diluted that they could rarely if ever interact with each other. If you have read elsewhere about the origin of life, you will have come across the phrase 'primitive soup' to describe the character of the terrestrial oceans before life appeared: a solution of mixed organic chemicals from which life is thought to have originated. The evocative metaphor was first used by Academician A. I. Oparin, a famous Soviet biochemist who published the first serious book on life's origins in the early 1920s. It is a respectable but unfortunate metaphor, for as soups go, the waters of the young Earth would have been very, very weak. However, there are ways in which the organic materials would have become concentrated. For one thing, they stick to clay and other minerals readily and would concentrate on exposed mineral surfaces. For another, they would accumulate on the shores and floors of drying-up lakes and seas, or even large pools. Localities in which the organic molecules would cluster together sufficiently to interact would therefore have existed, and would probably have been quite common.

I mentioned a moment ago that Miller–Urey substances include chemicals that today go to make up the nucleic acids referred to by the initials RNA and DNA. These substances are crucial to reproduction in all known organisms. To explain their peculiar importance, I must digress briefly into the way cells multiply.

I have mentioned in other chapters, and I shall repeat briefly, that a sort of genetic blueprint is passed from parent to offspring, in the form of a very long, coiled molecule known as DNA (the initials stand for deoxyribonucleic acid). Genes are particular stretches of DNA, and a whole DNA molecule might carry thousands of genes. It is a little like an audio tape: the information it carries is encoded on the tape – the genes might be compared to successive songs – and the cell has to have machinery to read the information and convert it into something useful. Crucial to the reading machinery are different chain-like molecules known collectively as RNA (ribonucleic acid). RNA molecules are involved at three stages. Members of one group of RNAs, aided by a protein, read bits of the information held by DNA – one gene, for example – and carry their message to the cell's protein-synthesising machinery. This is a structure which is made up of RNA (three distinct chains which are much longer than those of the messenger RNA) plus over 50 special kinds of protein; the combination makes up a sub-cellular particle which, to make new protein, sticks amino-acids together, ordered according to the message that the shorter chains of RNA have delivered. Those amino-acids, which come from food, are ferried to the protein-synthesising machinery by a third variety of RNA. The various proteins built up in this way then provide everything else the cell needs to grow and multiply.

I have simplified the story of how genetic information in DNA is used to make a living cell because the finer details do not matter just now; the point I wish to make is that, although DNA is tremendously important as the genetic library, it is RNA that does the heavy work. It transcribes the information stored in DNA, and it helps in two ways to translate it into proteins.

There are hundreds of different RNA molecules milling around in living cells, some carrying different messages, some transporting various amino-acids, in addition to the three large chains forming part of the protein-making machinery. But all RNA molecules are made up of smaller molecular units, called ribonucleotides, linked together chemically to form an RNA chain. The ribonucleotides themselves are compounds of the three components I mentioned earlier: inorganic matter called phosphate, a special sugar molecule called ribose, and a molecule called a base. Four different kinds of base occur in RNA, and permutations and combinations of these bases along the chain spell out the message that a messenger RNA carries from the DNA of genes to the protein factory. (For those of you who have looked at Chapter 17, two of the bases are the same as are found in the nucleotides of DNA, two are not.)

So, back to early planetary life. The sugar ribose and the four bases constituting RNA are commonly found among Miller–Urey compounds. Phosphates are common in nature, and Miller–Urey type experiments done in the presence of phosphates have actually yielded small amounts of ribonucleotides. So it is likely that ribonucleotides were present in the Earth's pre-life environment. DNA is in many ways rather like RNA. It, too, comprises chains of phosphate, sugar and base, but the sugar component is a different one called deoxyribose. It is not a Miller–Urey substance. So there is no reason to imagine that the building blocks of DNA were present in those ancient days. Therefore RNA is probably a more ancient substance than DNA.

Scientists who speculate on how life originated refer to the processes that led to its emergence as 'biopoesis'. I inform you about that word largely because it enables me to use a pleasingly appropriate term invented by N.W. Pirie, another substantial contributor to biopoetic theory. He calls such scientists 'biopoets'. It does not stretch the biopoetic imagination much to concede that ribonucleotides, suitably concentrated, might well have linked up in that distant era to form RNA-like chains. There is another substance which is universally found in living cells called by the initials ATP (I shall not bother with its full name here). The ATP molecule is the energy currency of the cell: it decomposes to release energy for growth, movement, maintenance and the other processes that constitute life, and the ultimate objective of food consumption and respiration is to generate ATP for this purpose (ATP has turned up in several chapters of this book, notably in Chapter 7). In living cells, ATP is constantly being formed from, and being converted back to, a ribonucleotide called AMP, of which every living cell has its quota. AMP, too, is a Miller–Urey compound. Biopoets in the 1960s and 1970s were much taken with the idea that RNA-like molecules were the precursors of life, and that AMP and RNA are among the vestiges of life's original biochemistry.

At that time, a flaw in the whole story was that there seemed to be no plausible way in which RNA could reproduce itself. In living cells, the actual syntheses leading to reproduction are performed by proteins, and proteins alone. The vast variety of molecules called enzymes, which bring about the chemical reactions that constitute being alive, are proteins. Enzymes are catalysts, very good and specialised ones; other major cell components, carbohydrate, DNA and RNA molecules, are not catalysts at all.

That is how matters stood until about a dozen years ago. Then a group of molecular biologists working at the University of Colorado, USA, discovered an RNA molecule in a protozoon which had a kind of enzymic activity. Given certain conditions, this RNA chain would chop a piece out of itself. (That might seem an odd thing to do, but it made sense in its context.) Since then a number of RNA molecules have been discovered which cleave bits out of themselves; they have been given the special name 'ribozymes', to emphasise their analogy to enzymes. They belong to the messenger type of RNA and they perform this self-splicing reaction to modify in transit the message that the molecule is carrying to the protein-making machinery.

Biopoets are today attracted by the idea that ribozymes are the relics of pre-life molecules. The scenario is that spontaneous linking up of ribonucleotide chains led to ribozyme-like molecules being formed, some types of which could assemble rather than cleave RNA chains (this is not an unreasonable suggestion; most biological reactions can be made to go backwards). Then they might come to reproduce each other, or, if two such molecules joined together, they might reproduce themselves. Out of the primal chemical mêlée a self-reproducing RNA molecule could thus have emerged spontaneously.

A self-reproducing length of RNA is a long way from an organised cell, but our modern biopoets like to think that a kind of RNA-based life emerged, as such self-reproducing molecules became more sophisticated, began to interact co-operatively, and sequestered themselves as cell-like units within a fatty membrane. In due course, they learned to string together those amino-acids which were lying around, and began to exploit the huge enzymic potential of proteins. Is this plausible? Biopoets would point to recently discovered examples of an amino-acid, a sugar or a ribonucleotide inhibiting the action of ribozymes, which implies that sites for interacting with such molecules exist on the ribozyme chain. If a ribozyme chain can interact with an amino-acid, say, it is not a big step, in chemical terms, to being able to modify the amino-acid so that it reacts with another amino-acid – which is a first step towards making a protein.

Once primitive RNA chains learned to make proteins that were also enzymes, they would have become the analogues of genes. In such a way, an 'RNA-world' is thought to have originated. Our present DNA-world emerged later, a way of life which became dominant because DNA was a more stable repository of the genetic library than RNA would have been. Once the RNA-world became established, and it may have fluttered into existence and been extinguished many times before that happened, its organisms would consume the organic matter of that 'primitive soup' and thus prevent spontaneous biopoesis from occurring again.

There is a precedent for RNA acting as a genetic library: it does so today in certain viruses, which are highly specialised parasitic microbes. However, their RNA is not self-reproducing: for multiplication, the virus recruits the DNA of the host it has infected.

A lot more detailed speculation and (as you may guess) argument concerning scenarios of this kind may be found in books and articles on biopoesis, but in the end it is all biopoetry. There is no hard evidence for anything more than the presence of biological-type chemicals, including ribonucleotides, in the pre-life environment. But the universality and the fundamental character of RNA and AMP (and some related compounds which I have not mentioned), together with the enzyme-like activity found in ribozymes, give it all a biochemical plausibility which earlier excursions into biopoetry lacked. And speculation is essential to scientific advance; as biopoetry goes, ideas have moved quite a way on from Oparin's primitive soup.

Perhaps I should add that, though I have described this scenario in the context of the Earth's history, a minority of biopoets takes the view that a period of under a billion years is too short a time for organised DNA-life to have arisen spontaneously. They prefer to imagine that events of this kind happened not here but somewhere else in the universe, and that life was seeded on this planet from elsewhere in a form comparable to today's viruses. This is an updating of a nineteenth-century concept termed 'panspermia', to the effect that the seeds of life are universal throughout the universe. Personally, I think our modern panspermic biopoets underestimate the rate at which a self-reproducing entity would evolve, once it had emerged, but who knows?

Now

Be that as it may – as my mother would have said – DNA-life is here and ours has become a very crowded planet. We share it with a vast variety of animals, plants and microbes, all crowding in on each other, shoving one another aside, competing to occupy the inhabitable space available. Among plants, competition for soil nutrients, sunlight, and often both, eliminates the less vigorous; among animals, nutrient supplies – food or prey – limit populations, and elaborate instincts and behaviour patterns have evolved concerned with the acquirement and defence of territory.

Microbes have like problems: when appropriate criteria of temperature, salinity, acidity, aeration, and so on have been met – a remarkably flexible set of criteria, as I showed in the early chapters of this book – then the nature and balance of a microbial population is determined by the local nutrient supply. Microbes which live in association with higher organisms, on skin or plant leaves, in animal intestines or root nodules, are on the whole lucky. They have a fairly reliable nutrient supply; they crowd the available space as much as they can, and live their lives in balance with it, needing to ensure only that they have adequate adhesive or invasive properties to stay where they need to be and grab what they can.

In the cold, hard outside world, in soil and water, life is rougher. Nutrient supplies fluctuate; famine and deprivation are the regular microbial experience, interspersed with occasional gluts of food, such as autumn leaf fall, the arrival of animal residues (excreta, regurgitations or corpses), or surges of nutrient brought about by weather or temperature change. In times of glut, the microbes which can multiply fastest (they are generally species of bacteria and micro-fungi) do so, and they grab most of the fresh nutrient. Space does not present them with a problem; they crowd around and on top of each other, forming colonies or masses of organisms, burgeoning until all the more readily assimilable components of the nutrient surge are converted into themselves. Sometimes a 'bloom', as it is called, of cyanobacteria, or of coloured photosynthetic bacteria, will cover a whole lake, turning it into a spectacular green or purple broth. But in time the fast-growing bacteria cease to multiply and, as famine re-establishes itself, they gradually die.

Their victory over their more slowly-responding cousins has been short-lived. Yet they have performed a valuable service to

the microbial community as a whole, because they have sequestered the nutrient, whatever it was, into themselves and then, as
they die, they release nutrient (albeit of a different kind, derived
from their disintegrating remains) at a much slower pace. Part of
that nutrient might be cannibalised by a few surviving brethren
belonging to their own species, which thus continue to maintain
themselves ready for the next glut, but most of it goes to their
slow-coach cousins, who are better scavengers and who then
actually flourish.

This is the stage at which space becomes important. The slow-
growing types, although very good at scavenging scarce nutrients,
find it advantageous to occupy as large a share of the available
space as they can, because that way they can grab nutrient from
a greater volume of territory.

Microbes, unlike people, do not actively fight each other for
space. But many defend their territory passively, simply by being
there. A good scavenger removes nutrients from its immediate
neighbourhood as fast as those nutrients arrive, so potential competitors are starved out and do not get a foothold. In addition,
some of the end products of microbial metabolism are toxic to
other organisms. An especially clear example is provided by a
group called the sulphate-reducing bacteria (they have appeared
before and were featured in Chapter 7). These are anaerobes –
they do not use gaseous oxygen and cannot grow in air – and
their respiration leads to the formation of hydrogen sulphide, a
smelly gas that is poisonous to most air-breathing microbes (and
to higher organisms as well). However, it positively improves the
scene for the sulphate reducers, because it reacts spontaneously
with oxygen and so helps to keep it out of their neighbourhood.
It also kills off any air-breathing neighbours, thus enhancing the
supply of nutrients. That kind of passive territorialism, in which
a species alters its environment to favour itself, is not uncommon
in the microbial world (nor, if you think about it, among higher
organisms). As further examples (gleaned from Chapter 6), acid-
producing microbes, being inherently more tolerant of acidity
than many, favour themselves by making the environment acid,
and alkali-generating types help themselves by making their habitat alkaline. But in all such instances evolution has thrown up a

plethora of other microbial species which are happy to go along with the change: other anaerobes flourish alongside sulphate reducers and some even come to depend on the sulphide they make; micro-fungi and yeasts like a spot of acidity; a special micro-flora of alkali-tolerant microbes exists. And in every instance there is a limit: too much sulphide inhibits even a sulphate reducer, and too much acid or alkali can put a stop to even the toughest of tolerant microbes.

Micro-chemical warfare

Competition for nutrients, and making toxic products, are passive, and only moderately effective, ways of excluding other microbes from a given microbial niche. For several decades, microbiologists have wondered: does a more active form of territorial strife exist among microbes in nature? The idea arises because a large number of microbes produce substances which inhibit the growth of other microbes.

Such substances are especially common among a group of soil microbes called streptomycetes. These are a fairly sophisticated group of bacteria. They grow slowly, usually as a filament, and this enables them to cover a lot of space when seeking food: they are good scavengers. And unlike the great majority of bacteria, they show a primitive kind of differentiation. If the environment becomes especially adverse, certain of the filaments grow upwards, like hairs, and at the tips of these hairs they form spores, which are dormant bodies which hardly metabolise at all and which are resistant to stress. These spores will be dispersed by wind, rain, animals and so on when the filament dies. Before things get so bad that they form spores, the bacteria produce chemicals which prevent the multiplication of other species of microbe. These chemicals are very active, in the sense that tiny concentrations inhibit susceptible bacteria. The collective name for them will be familiar to most readers: they are called antibiotics.

As well as streptomycetes, moulds (the micro-fungi which grow on bread and cheese as well as inhabiting soil) make antibiotics. Moulds also form spores. Members of a genus of bacteria called *Bacillus* often produce antibiotics, too, and again they form spores (though in a less elaborate way: individual cells simply turn into single spores if stressed). Ability to make antibiotics seems vaguely to parallel ability to form spores, but not always: many spore-formers do not form antibiotics, and a few non-spore-formers do.

Actually, ability to make anti-microbial substances is not restricted to microbes; certain species of frog, for example, secrete chemicals which prevent bacteria growing on their skin (which would otherwise be an invitingly damp habitat), and most animals have anti-bacterial substances in bodily secretions such as sweat and tears. But the name 'antibiotic' is usually reserved for chemicals which are released by microbes and which are detrimental to other microbes.

Antibiotics were brought into medical use in the 1940s, and during the 1950s it seemed as though new ones were introduced every few months, but their history goes back another decade. Probably the most famous is penicillin, the first to be discovered, which is made by a common mould called *Penicillium*. Penicillin is indeed a 'wonder drug': it has revolutionised medical practice during the last 40 or so years, curing and virtually eliminating several bacterial diseases, and it remains extremely valuable,

The story of the discovery and development of penicillin has been told often, but I shall repeat it briefly here because, being a tale of untidiness, accident, improvisation, drama and exploitation, it provides an instructive vignette of the manner in which science and its applications muddle along.

In 1929, Alexander (later Sir Alexander) Fleming, a medical bacteriologist, was working in his laboratory at St. Mary's Hospital, Paddington (London). His job was to test the effectiveness of disinfectants on various disease-causing bacteria. He was not a tidy worker; his habit of leaving his test cultures on the bench when he had finished with them, instead of sterilising and discarding them, was (and still is) frowned upon by experts. Rightly, because they can become muddled, or become contaminated by unwanted airborne or dust-borne microbes, or simply be knocked over by accident, and spread infectious material about. In fact, the second of these hazards happened fairly often to Fleming's cultures. He grew his cultures in a conventional way: in a flat dish with glass cover, containing a sort of jellified meat broth on the surface of which bacteria could multiply and form colonies. On one occasion, a dish in which he had grown colonies of staphylococci (bacteria which cause boils, among several nastier ailments) had lain on his bench for some weeks when he noticed that a spore of a mould had somehow got in, and was growing as a colony, spreading outwards from near one edge. This was not an unusual sight to Fleming. What was peculiar, however, was that, where the mould colony grew, and in a zone stretching about 2 centimetres ahead of its advancing front, the colonies of staphylococcus were disintegrating and the bacteria disappearing.

Many bacteriologists would have discarded the dish, but Fleming was intrigued. Obviously the mould was releasing something into the jelly which killed the bacteria. So he isolated a culture of the mould, which proved to be *Penicillium*, showed that it would repeat its trick of dissolving staphylococcus colonies, and made some collaborative attempts to isolate the substance, which he named penicillin. He failed. So he stored the mould culture, published an account of its activity, and got on with his regular work.

Half a dozen years later a scientist from central Europe called Ernst Chain, a refugee from Hilter's persecution (and also a more experienced biochemist than Fleming), was working at Oxford on an enzyme from mammals that dissolved bacteria. He came across Fleming's paper, which seemed to suggest that a mould made something of the sort. With the agreement of the head of his laboratory, Howard Florey, he had another try at isolating penicillin, and by 1940 – by which time the Second World War had started – he and his colleagues had obtained a small amount, a few milligrammes, of material which he thought must be almost pure because it was so highly active against staphylococci. Opportunity for a clinical trial came all too soon: a policeman in the Radcliffe Hospital had acute staphylococcal blood poisoning for which there was at that time no cure. A course of treatment using the limited amount of material available was started and a spectacular recovery got under way – but tragically the supply ran out too soon: the policeman relapsed and died.

What penicillin actually was chemically remained an enigma, but Chain and Florey had already learned enough about its medical promise to set about growing the mould on a large scale so as to extract substantial amounts of the substance from its culture fluid. Large fermenting vats were not abundant in wartime, so milk churns had been converted for their pilot work. Because of its potential value to the military, Britain's wartime ally, the USA, agreed to develop larger scale production of penicillin. The Americans improved both production techniques and the culture medium. Pharmaceutical industries became involved and the improvements were patented so that, after the war had ended, the British (to Chain's fury) were obliged to pay royalties to the USA for the right to manufacture penicillin efficiently.

Pure penicillin proved to be some 100 times as active as the material used on the unfortunate policeman: that material had been far from pure. It also cured many bacterial diseases other than those caused by staphylococci. In later years, by genetic manipulation and attention to the culture conditions, pharmaceutical companies obtained strains of *Penicillium* which were able to make two or three hundred times as much penicillin as Fleming's original population.

I gloss over the consequent tales of secrecy among researchers in different laboratories, of broken security and industrial espionage, of black markets in dubious penicillin in the immediate postwar period. Penicillin became a reasonably cheap commodity, while remaining a highly profitable product for pharmaceutical companies and, searching for new antibiotics, their research departments devoted years to screening microbes, and their products, for anti-microbial activity. Well over a thousand new antibiotics turned up, and by the 1990s about a hundred were in use clinically or in agriculture. Several chemically modified forms of penicillin, with improved properties for particular diseases, have been developed too. Fleming, Chain and Florey shared a Nobel prize for their achievement.

It was a vignette of scientific advance, as I said. The story has several morals but I shall emphasise only one. To Fleming, penicillin was a scientific curiosity. He had no thought, as he later acknowledged, that it might be of medical value. And both Florey and Chain have stated that medical use was not uppermost in their minds: both regarded it primarily as an interesting research problem. Most advances in the applications of science, from electricity to atomic energy, have arisen from comparable disinterested scientific curiosity: it is usually the trimmings that are foreseen and planned.

Back to this chapter's theme. The chemical structure of the penicillin molecule was worked out by the late 1940s. It proved to be quite a simple structure, but unlike any other known biochemical substance. Today the structures of most of those other thousand antibiotics are known, and nearly all are peculiar: small organic molecules with unusual chemical structures. Like penicillin, they usually turn up in laboratory cultures towards or after the end of multiplication, when the microbial population is to some extent stressed. One wonders why.

For the last few decades, microbiologists have believed – no, I should say half-believed – that antibiotics are weapons in a sort of chemical warfare waged by one species of microbe against its competitors. For example, streptomycetes, which (I remind you) are filamentous soil bacteria, can, if they generate antibiotics at all, release enough to stop sensitive bacteria from growing in a zone about 10 micrometres round the filament (which itself is about 1 μm thick). And a variety of anti-bacterial substances called 'bacteriocins' are made by certain strains of bacteria which kill other strains of the same species: a positively human type of warfare.

These facts might seem to support the idea of chemical warfare, but it has major flaws. For one thing, most soil bacteria are not sensitive to antibiotics, and even among those that show sensitivity, populations soon acquire resistance when exposed to a low, gradually increasing concentration of antibiotic – which is just what a wild streptomycete or mould would generate. That resistance results from a change in the genetic apparatus of the cell: once acquired, it is handed on from generation to generation. Indeed, soil microbiologists sometimes use the varieties of antibiotic to which soil bacteria are naturally resistant as a sort of

'fingerprint': a pattern of properties which enables them to identify local strains. (It works because those from one habitat may
have encountered, and learned to resist, a different range from
those from another habitat. Incidentally, if a scientist wants to
add a strain of bacteria to a natural environment, and be able to
recognise its progeny later on, he or she will make it resistant
to an uncommon antibiotic in the laboratory first.) And there are
other flaws: in most natural ecosystems, varieties of microbe
which do not form antibiotics are many times more abundant
than are their antibiotic-producing brethren. Finally, many of
the known antibiotics, such as penicillin, are most active against
microbes which the producer would rarely if ever encounter in
soil, and which would be trivial competitors if they did appear
there.

Bacteriocins might seem to have some plausibility as chemical
weapons, and some may be, but one of the few whose mode of
action has been worked out, a 'colicin' formed by a bacterium
which inhabits mammalian guts called *Escherichia coli*, proves to
be not a weapon but an aid to scavenging: it is a chemical which
enables its possessor to scavenge a particular nutrient especially
effectively. It inhibits competing strains of *E. coli* by starving them
of iron.

In short, antibiotics would seem to be short-lived and unreliable instruments of aggression. Even as defensive weapons
– protecting cells during spore formation has been proposed as
their natural function – it is not clear what threat antibiotics would
repel. Occasions on which an antibiotic might be useful in the
struggle for space would seem to be so rare that one is forced to
the conclusion that they are not primarily territorial weapons, and
that if they sometimes work that way, they do so fortuitously, like
the sulphide or acidity produced by other bacteria.

Then why do these enigmatic substances appear at all? Any
explanation must take into account some other features, too. The
chemical structures of most antibiotics have been elucidated and,
in a biochemical sense, nearly all are peculiar. Bacteriocins are
proteins, but the majority of the rest are small organic molecules
with no obvious relationship to the biochemicals encountered in
the main stream of cell metabolism; they tend to turn up in

cultures towards, or just after, the end of multiplication. Rarely do they positively kill bacteria; some do, but more often they interfere with their victims' growth and multiplication. Penicillin, for example, prevents susceptible bacteria from making proper cell walls, so it does not affect non-growing bacteria. If they start to grow, however, their walls fail to keep up and the cells burst (that is why Fleming's staphylococci vanished).

Julian Davies, a contemporary expert on antibiotics and their related genetics, has recently made a novel proposal. He pointed out that a great many antibiotics (though not penicillin) act by interfering with the functioning of RNA. For example, streptomycin, the second antibiotic to be discovered, interferes with growth by tangling with the protein-synthesising machinery of which I wrote earlier. Exactly how it works is still not clear; a streptomycin molecule sticks very tightly to the protein-synthesising particle, and it probably adheres to a ribonucleotide in the RNA part of the machinery, making an important but subtle difference which incapacitates susceptible cells. I use the term subtle because very occasionally susceptible bacteria mutate and become resistant to streptomycin, and if this happens the machinery works even with streptomycin stuck to it. Moreover, in a special, rarer, kind of mutant, the whole machinery changes so that it does not work *unless* streptomycin is there. To put the point differently, susceptible bacteria can mutate so that they positively require a fix of streptomycin to multiply!

There are numerous other antibiotics which, like streptomycin, tangle with RNA. Mutants which resist them are known, and mutants which change from being susceptible to requiring the antibiotic have been reported, too.

Why should substances which interact so drastically with RNA have arisen? They even present the organisms which make them with a problem: they have to alter them chemically, or export than quickly, or they would gum up their own RNA, so to speak. Davies suggested that they exist in cells because they are left-overs from the distant era of RNA-based life. Just as the human appendix is a relic of an earlier period in our evolution, so anti-biotics – or at least many of them – are relics of an earlier phase of life's development. The idea is that they, or substances very like them, were formed by Miller–Urey reactions alongside the pristine ribonucleotides. As these evolved into self-replicating RNA-life, the antibiotics-to-be helped in transmitting genetic information from parent to daughter. To-day's DNA-life needs regulator substances, chemicals which control, switch on or off, the various biochemical activities which DNA master-minds. These regulators are almost always proteins. And in a less conventional sense, DNA itself is a regulator, in the sense that, like a template, it dictates the chemical structure of the product it specifies. RNA-life would have needed regulators and templates, too, and they would need to have been simpler molecules, ones which worked by interacting with RNA in reversible ways, perhaps acting as templates on which amino-acids were strung together. Antibiotic molecules, Davies suggests, once did that sort of thing (he favours a template role) and, though the functions of RNA have changed in our DNA world, the sorts of site where they would have interacted still exist on RNA molecules.

But there is still a flaw. Evolutionary relics do not just hang about; over successive generations they vanish under the pressure of natural selection, either disappearing altogether (like the human tail) or transforming into something new (as a fin became a leg). Why should seemingly useless relics of RNA-life persist today? The response is that they are not useless; that they still have functions in cells, but at much lower concentrations. And indeed one can adduce evidence for a variety of such secondary effects: low, sub-inhibitory concentrations of a wide variety of antibiotics have been recorded as being able to stimulate steps in the operation of the normal genetic machinery, as well as accelerating cell growth and interrupting ribozyme action. The drastic antibiotic effects that enabled scientists to discover them are due to overdoses.

One beauty of Davies's idea is that it rationalises the existence of a wide variety of other substances formed by moulds, streptomycetes and other microbes, and even by higher fungi, which are not antibiotics. They are called secondary metabolites. They are substances with small molecules of unusual structures, of no obvious use to the organisms that make them. Yet quite often they prove to be biologically active – usually toxic in some way – to others. They appear most often when growth and multiplication have slowed down or ceased. They, too, might be evolutionary left-overs, conserved because of some physiological function yet to be discovered. Scientists noticed the antibiotics first, because they were useful, but even in 1950, when the structure of the penicillin molecule was finally resolved, more than 100 odd-ball molecules produced by moulds had been described in the scientific literature, having been discovered and purified by organic chemists in the previous twenty-odd years. (Disappointingly for those chemists, a belated testing of them all revealed only one substance with antibiotic activity, and that was too poisonous for medical use.)

Are antibiotics and secondary metabolites present-day relics of extremely ancient participants in long-obsolete genetic and metabolic processes, left over from distant eras of biological evolution but still useful? That is a mouthful of a question – but it is an attractive thought, and a considerable improvement on the alternative idea that they are useless junk. Perhaps it will be a fruitful thought, too, and the further study of what antibiotics do will teach us more about the earliest stages of terrestrial life.

Well, evolutionary relics or not, antibiotics are of little strategic or tactical use in a microbial battle for a private space. They work marvellously for mankind, of course, but only because we extract them and use them wholly out of their context. Among microbes, an antibiotic may briefly damage or dismay the unsuspecting neighbour but, as chemical weapons, in the long run antibiotics do not work.

Grab what you can and shove others aside seems to be the only microbial strategy for acquiring *lebensraum*. It is an unappealing feature, I fear, but one which permeates biology even to the most highly evolved creatures. Just look about you . . .

15

Company

The survival of the fittest

C harles Darwin, the great nineteenth-century biologist, is
widely thought to have originated the idea of biological evolution. He did not. The concept had been around for a couple of
generations before him. In the eighteenth century, naturalists and
geologists had noticed that species change over time, that some
die out and others improve and flourish. Clearly, they reasoned,
plants and animals evolved, but they could not think why or how;
they tended to regard evolution as yet another manifestation of
the wondrous works of The Creator. What Darwin did, and this
was his tremendous contribution to scientific thought, was to
explain how evolution worked: he introduced the concept of natural selection. He also, making use of his own meticulous observa-
tions made over some 30 years, presented overwhelming evidence

that he was right, to the indignation of the established church (and, in truth, somewhat to his own embarrassment, because his wife and many of his family were staunch Christians and believed literally in a moment of divine creation).

What natural selection said, in essence, was this. It is obvious that sexual reproduction leads to mixing of inheritable characters. Therefore generations of offspring differ in various ways from their parents, in particular in the extent to which they are adjusted to their surroundings. Among animals, for instance, some may find food more efficiently than others, may run faster from predators, may hunt more cleverly, may be more attractive to mates and so on. Those best fitted to their habitats are likely to live longer and breed more offspring than their less competent brethren. Therefore the favourable hereditary qualities from which they benefited will tend to be conserved by the species, and to become spread among successive generations. Gradually species will change as they adapt more effectively to their habitats. Less well-adapted species may be competed out altogether; over a long time, accumulated inheritable changes will lead to new species emerging from previous types.

In effect, the habitat will have exerted selection, albeit passively, on successive generations of the species which inhabit it.

For centuries artificial selection had been used by animal breeders, such as pigeon fanciers and cattle breeders. Darwin showed how selection could, indeed must, occur spontaneously in nature, and lead to change within species as well as to the origin of new species. In the mid nineteenth century the idea that the nature of the living world was determined by natural selection was revolutionary, especially when Darwin followed his logic through and pointed out that the ancestors of Man must have been ape-like creatures. Then indeed was there uproar among the clerics! The fuss approached that generated by Galileo's contention, over two centuries earlier, that the Earth was not the centre of the solar system. (Happily times had changed: Darwin was not incarcerated for heresy.)

Darwinism, as it came to be called, excited not only biologists and geologists, but also philosophers, politicians and writers, as well as much of the lay public. People's ideas of the world around them, and their place in it, were questioned and transformed. 'Nature red in tooth and claw', wrote Alfred, Lord Tennyson, the poet. 'The survival of the fittest', declaimed the philosopher Herbert Spencer.

The political and social fall-out of this new biological thinking was sometimes bizarre. Atheists and free-thinkers were delighted, maintaining that Darwinism removed the need to believe in God at all. And the thought that competition, the struggle to out-do your fellows, was a Law of Nature appealed to many successful bourgeois as seeming to justify the inequalities inherent in Victorian Britain – and throughout the world. On the other hand, the growing Socialist movement saw Darwinism as confirmation that Man could, and should, evolve towards a wiser, more just and humane, society. In short, Darwinism became all things to all men – except to mainstream Judaeo-Christians.

For as Darwin himself had foreseen, his ideas were not so much misunderstood as understood selectively. Indeed, being peaky in health and of mild disposition, Darwin had so feared the uproar and confusion his ideas would generate that he delayed publishing them for some 20 years, and did so only when a young fellow naturalist, Alfred Wallace, reached similar conclusions independently. They published simultaneously.

Living together

In this century, much of that furore has died down. Today Darwin's theory of evolution is as central to biology as Dalton's atomic theory is to chemistry. Both have been subject to incessant questioning and modification as knowledge has progressed, and today both are basically unassailable; only ignoramuses and cranks challenge either. In the words of the late Sir Julian Huxley, a great twentieth-century biologist whose work and writings did much to place Darwinism in its central position, evolution is 'one of the great liberating concepts of science'.

And we now know a lot more about it. Darwinism works very well for much of biology, but there are detailed instances where it seems not to work, or to operate in a convoluted manner. We also know that co-operation has been as important as competition in evolution. Moreover, we have a far clearer insight into the ways in which species influence the selection pressure which operates upon them: how they alter their habitat, often adjusting it to suit themselves. Nature changes species, and species change nature: both the living world and the inanimate world evolve in partnership.

Co-operative interaction becomes especially clear when you look at the global picture, because living things all depend on each other for existence. Consider, for example, the way in which animals cannot do without plants.

Animals need plants for food, and hence for their existence. In practice they may consume plants at one or two removes, by eating herbivorous animals rather than plants themselves, but if there were no plants there would be no animals – or none of the kind we know. Plants, aided by sunlight, make carbon dioxide from the air into the organic matter which constitutes themselves (the process called photosynthesis), and that process sustains both themselves and the animal world. But there is more to the story. Animals need oxygen to breathe, and green plants make it, as a by-product of photosynthesis. Again, if they did not, we should not be here. There was a time, 2 or 3 billion years ago, when there was no free oxygen in the Earth's atmosphere. Nevertheless a menagerie of microbes had evolved and, among these, were bacteria that could use sunlight to convert carbon dioxide into organic matter. These early types did not make oxygen as a by-product. However, the kind of photosynthesis which produces oxygen is biochemically more efficient than the more primitive ways, and in due course a sub-group of photosynthetic bacteria learned the trick. Thus they became the distant ancestors of today's green plants (though in a convoluted way, as I shall tell later). Duly plants appeared on the planet – about half a billion years ago – and by producing oxygen over successive millions of years, plants transformed the Earth's atmosphere into what we have today (give or take a spot of man-made pollution).

Thus plants radically altered the kind of natural selection that operated upon planetary life, including themselves, as well as setting things up for the emergence of animals.

The tie-up between plants and animals does not end there. The carbon dioxide which plants use comes from the air, but supplies have never run out because living things give it back again. Respiration (which amounts to controlled burning of organic food by plants, animals and microbes) returns some carbon dioxide, but by far the greatest amount is returned through the decomposition, putrefaction and decay of dead or excreted organic matter. And the agents responsible for the latter processes are bacteria and fungi: almost exclusively microbes.

These processes go on today. Every year carbon atoms equivalent to something in excess of 400 billion tonnes of carbon dioxide move from organic to inorganic form through biological activities in the air, waters and land mass of our planet. Biologists call this turnover of carbon compounds the carbon cycle, and its details are much more subtle and complex than my outline would suggest. (They have become especially cogent as this century ends, because mankind's use of fossil fuels has unbalanced the cycle, leading to more carbon dioxide in the air and the prospect of global warming.) However, the message relevant to evolution is that, as far as carbon compounds are concerned, plants, animals and microbes have to co-operate; they could not live without each other.

Carbon compounds, as organic matter, are the major components of living things. But carbon compounds always contain other elements. Hydrogen and oxygen are usually present, nitrogen is often there too, sulphur is important but less common; then come phosphorus, iron and a few minor, but far from unimportant, elements. The first four, and probably most of the rest, undergo biological cycles analogous to the carbon cycle, and in some of these cycles microbes are transcendentally important. The nitrogen cycle is an excellent example. Every living thing contains 10 to 15 per cent of nitrogen (in chemical combination, of course), but neither animals, plants, nor yet most microbes, can make use of the vast reserves of nitrogen gas which constitute four fifths of our atmosphere. However, a specialised group of bacteria called nitrogen-fixing bacteria (which appeared in

Chapters 10 and 11) take nitrogen gas from the atmosphere and make it into a form that plants can use as fertiliser. From plants the nitrogen goes, as organic nitrogen compounds, to animals; then it is returned to soil and water through their excretions, and death, passed on as a result of microbial decay processes. Though much nitrogen gets recycled into plants direct – dung is jolly good fertiliser – some 10% annually is returned to the atmosphere as nitrogen gas (by a different group of bacteria). Happily the nitrogen-fixing bacteria recover it again. In the matter of nitrogen, all higher organisms depend on this special group of bacteria for their continued survival.

Oxygen, hydrogen and sulphur undergo comparable cyclical transformations in and out of biological matter, though the cycles differ in detail. In all of them, bacteria are particularly important: they conduct some quite exotic biological reactions and have already represented some of the outer reaches of life in earlier chapters of this book. But for present purposes the message is that the whole of terrestrial life is a network, a complex of web-like interdependent systems. Animals, plants and microbes cannot manage without each other and, collectively, they make the world what it is today, sustaining its atmosphere and, in numerous ways that I cannot go into now, maintaining the character of its seas and soils, even its weather and temperature.

Living close together

Co-operation of the kind I have just described is, of course, a fortuitous collaboration, as mindless as the converse struggle for survival. And within life's overall interactive structure, the more competitive aspects of natural selection operate unimpaired, as species battle for suitable niches within the global environment.

But as if to balance the struggle for existence there are many instances of seemingly more overt co-operation. Perhaps the most familiar example would be plants, which feed pollinating insects and have their pollen distributed in return, and also provide havens and food for birds and climbing mammals, who distribute their seeds in return. Other examples include oxbirds, which rhinoceroses and cattle tolerate on their backs because they pick off and eat parasites; aphids, which ants nurture and graze for their sugary secretion; pilot fish, which warn whales of approaching enemies. These are the stuff of general biology textbooks: the world of higher organisms is fraught with examples of associations among disparate creatures which are beneficial to both parties (if not always in equal measure). During evolution some such associations have become so beneficial to one or other party that they are now a virtual necessity.

At this stage I must set aside (I choose my words carefully) a few technical terms. An habitual association between two (or a few) species is known technically as 'a symbiosis'. The terms 'a consortium', 'mutualism' or 'commensalism' are also used and they have different, though overlapping, meanings. Regrettably, some imprecision has crept into the way they are used by zoologists, botanists and microbiologists as the case may be. Therefore, though all the associations I shall write about in this chapter are in a broad sense symbioses, my account will lose nothing, and be simpler, if I make no attempt to be more precise.

Many of the associations which microbes form with higher organisms have become obligatory. Not all these associations are mutually beneficial (for example, microbes can cause disease in plants and animals), but most are. No-one knows the real figure, but at a guess I should say that half of the microbial world lives in direct association with a higher organism of one kind or another, and the subject of symbiotic associations among microbes and between microbes and higher organisms is vast. I shall have to omit many fascinating by-ways: the light-emitting bacteria which inhabit the luminous organs of deep-sea fish; the *Penicillium*-like mould which is cultured and farmed underground by ambrosia beetles; the micro-algae which colour marine worms, polyps and clams, and help them feed. The topic is almost inexhaustible; here I shall offer only glimpses, which will illustrate the surprising levels of intimacy that have evolved over the aeons.

Every gardener knows (and hates) couch grass. For non-gardeners, the name covers several kinds of coarse, wild grass that grow long underground roots and can take over the lawn and persistently invade the flower beds. Mediterranean and sub-tropical lands suffer from a hardy couch grass called *Paspalum notatum* (Britain is a mite too cold for it, though it can be found in the Southern counties). It flourishes almost anywhere, even in very infertile soils, and part of the reason is that it recruits a microbial friend to help. A bacterium called *Azotobacter paspali* makes a habit of living on the surface of this particular grass's roots, and it is one of the nitrogen-fixing bacteria that I mentioned in the last section: it takes nitrogen from the air and fertilises the soil in the immediate neighbourhood of the roots, benefiting the plant. In return, the plant's roots leak nutrients which the bacteria can consume. And there is more to the association than just shared nutrients. The enzymes responsible for nitrogen fixation are very easily damaged by oxygen, so the process is very sensitive to the presence of oxygen from air, and there is air even deep in soil. When the association is working well, the plant's roots, and their attendant bacteria, become coated with a sheath of starchy material which helps to restrict access of oxygen to the bacteria and so makes it easier for them to fix nitrogen. The grass helps the bacteria, and the bacteria help the grass.

Dr Johanna Döbereiner, a notable Brazilian microbiologist, discovered this association in a way that high-lights the oxygen problem. In the mid 1960s she published a short paper reporting Azotobacters around *Paspalum* roots, and claiming that they were beneficial to the plant. However, quite sensitive tests for nitrogen fixation were becoming available about that time and when these were applied to the roots, with their bacteria, they gave negative results: there was no evidence that the azotobacters were actually fixing any nitrogen at all. But as the 1960s progressed, bio-chemical research revealed the oxygen sensitivity of nitrogen fix-ation. Learning of these developments, on a working trip to Rothamsted Experimental Station in Britain in the early 1970s, Dr Döbereiner retested *Paspalum* roots, not in air but in a gas mixture with a low oxygen concentration. The tests were positive; her earlier research was vindicated and a new class of plant–bacterial associations was discovered.

Many comparable associations between nitrogen-fixing bac-teria and plant roots, including those of some familiar weeds, have since been discovered, though few are as effective as hers. All are casual in the sense that both plant and microbe can, and often do, get along adequately without each other, but they manage better together.

A second example – or set of examples – is provided by the many beneficial associations which exist between animals and microbes. Consider humans. We carry a positive zoo of bacteria on our skin, and in our mouths, throats and intestines. These creatures are not just harmless fellow travellers, they actually do us good by occupying niches where more harmful bacteria could become established – as we learn when we injure ourselves or become sick. Those that inhabit the lower gut – the part beyond the stomach – are particularly valuable because they make vitamins inside us. My one-time research supervisor, the late Dr Donald Woods, used to tell a story of how these bacteria fouled up nutritional studies on humans in the 1940s. It was the time of the Second World War, and the US military authorities wanted to study the sort of vitamin deficiencies to which their forces might be exposed in the Far Eastern theatre of war. So they obtained volunteers who agreed to become vitamin-deficient by living on diets lacking vitamins of the so-called B group. Weeks went past and the volunteers remained in rude health. At that time forms of the sulphonamide drugs were coming into use for the treatment of intestinal infections (they were never much good, actually), and someone – it may have been Woods himself, because he was an expert on these drugs – had the idea of administering a dose of the drug to the volunteers, to alter their intestinal microflora. It worked. Within days the group began to show the expected vitamin deficiencies and the research could proceed.

Message: normal intestinal microbes are invaluable to both human and animal health.

Bacteria also form mutually beneficial associations among themselves. One such was thought to be a single, though remarkable, bacterium when it was discovered in 1906, so it rejoices in its own name: *Chlorochromatium*. It is rare, but can sometimes be found among the various exotic bacteria which live in the sulphide-rich waters of sulphur springs. The 'organism' is large but uniform in size and green in colour; it needs light to grow because it photosynthesises. Under a good microscope it can be seen to consist of several green cells forming a bundle which surrounds a central, colourless rod-like cell, which has a tail enabling it to swim about. *Chlorochromatium* was especially interesting to microbiologists because such anatomical complexity was most unusual for a bacterium. However, it has now been recognised as a cluster of two distinct kinds of sulphur bacteria; the rod is a non-photosynthesising organism and its green companions do photosynthesise. For those who have read Chapters 7 and 8, they are thought to be a central sulphate reducer which generates sulphide from sulphate, which product is used by the surrounding green sulphide oxidisers which convert it, aided by light, back to sulphate. But do not bother to look up those chapters; the important point for the present is that their nutritional needs are complementary: each partner makes a nutrient which the other needs. The two kinds of bacterium multiply in unison, which is why the cluster was thought to be an individual microbe. The component organisms have not been persuaded to grow separately.

Living very close together

When biological co-operation becomes especially advantageous to both parties, natural selection exerts its usual pressure and the two kinds of organism not only live their whole lives together but evolve so as to change their anatomy or physiology in ways that favour the association.

A plant–microbe association of a close kind, familiar to most people because of its agricultural and horticultural importance, is that which clover and related leguminous plants (peas, beans, lucerne, lupins, etc.) form with strains of nitrogen-fixing bacteria from soil called *Rhizobium*. The association can fix nitrogen 10 to 30 times more effectively than the *Paspalum* association I wrote of earlier – hence its practical utility in crop rotation, for example. The reason for this high efficiency is that the plants provide special niches for their resident microbes, in the form of nodules which grow on the roots. This is how it happens. The growing roots of a young legume – clover, for example – will release into the soil small amounts of a chemical which acts as an attractant to the right kind of rhizobium. The rhizobia sense this (I wrote about bacterial senses in Chapter 13) and swim towards its source. Once they reach the right kind of growing root, they infect it, moving in among the cells until they reach one which they invade and colonise. The plant is, figuratively speaking, waiting for this. As the rhizobia multiply, the cell swells and the plant grows a nodule round it, as well as secreting special substances (a red protein like the haemoglobin of animal blood, for instance) which assist nitrogen fixation by keeping the oxygen supply low. The rhizobia gradually become dormant within their nodule, and devote their lives to fixing nitrogen, releasing most of the product to the plant, and being fed in return by the plant's vascular system. Rhizobia are released to find a new plant when their host dies. Some rhizobia are unable to fix nitrogen at all until they have settled into a nodule.

In the leguminous nodule, and in comparable close associations between plants and nitrogen-fixing bacteria, evolution has thrown up, so to speak, a sophisticated and rewarding association for both organisms, to which both partners have become adjusted.

Turning to animals, perhaps you think that sheep and cattle live on grass? If you have read Chapter 8 you will know that they do not; they live on butyric acid, spiced with some other acids and a digest of bacteria. Sheep and cattle and, indeed, most herbivores, are walking cultures of the microbes that live in the first stomach, the so-called rumen, of the animal. The fluid in the rumen is a dense 'soup' of microbes, mostly bacteria but including some microscopic animal-like creatures called protozoa. Most of this microbial menagerie is finely adapted to its cosy home: few can survive outside the rumen except in special laboratory cultures. There is virtually no oxygen in a rumen and they are adapted to an air-free way of life (I discussed bacteria which live without air in Chapter 8; animal-like creatures which live without air are very rare, except among the sorts of protozoa which live in rumens). The bacteria digest the cellulose of plants on which the herbivores graze – the animals themselves have no way of doing this – forming substances called fatty acids which the host animal can absorb. The principal fatty acid has a buttery smell and is called butyric acid; as I said, it is effectively the principal food of the herbivores, though they also digest some of the microbial cells which pass to the hind gut. A by-product of the cellulose breakdown in the rumen is gas: a mixture of carbon dioxide, methane and a little hydrogen (a flammable mixture, actually). The gas is responsible for the monstrous belches which ruminants such as cattle and sheep regularly produce (but if you have not observed the eructation of a camel you do not know what a real belch is. And while I am on these recondite matters, horses and other hoofed mammals have rather a similar mode of nutrition, but they house their microbial communities in their hind guts, so the gas comes out at the far end).

Herbivores have become completely dependent on their intestinal bacteria, but the protozoa probably do little for their hosts, only claiming a tithe of their food source. Since the majority of rumen microbes cannot survive outside the animal, the association has become mutually indispensable.

The rumen association has been known for many years; it enables ruminants to use a foodstuff – cellulose – which would otherwise be inaccessible to them. In the last decade associations have been discovered among marine invertebrates which also enable them to use otherwise inaccessible substrates, not cellulose this time but inorganic sulphides.

Volcanic springs exist on the ocean floor as well as on land, and bacteria, exploiting nutrients in the emerging spring water, become the primary food source for local populations of sessile worms, clams and occasionally crabs. (The hot, highly compressed geothermal springs of the Pacific Ocean, and their peculiar flora, cropped up in Chapters 2 and 4). The principal water-borne substance which these bacteria exploit is not a normal nutrient but a mineral: hydrogen sulphide. (A derivative, thiosulphate, is also used.) I described bacteria which live on minerals in Chapter 8; here I shall repeat that among the group called sulphur bacteria are types which cause sulphide (and its derivatives) to react with oxygen, a process which provides them with energy, and they use that energy to make organic matter from the carbon dioxide dissolved in the surrounding water. In something of the way that plants use solar energy to make carbon dioxide into organic matter, so these bacteria use chemical energy from oxidising sulphide to do a similar thing.

About a decade ago, examination of the worms and clams which live around the deep Pacific hot springs showed that they did not graze on the local bacteria; instead they had adopted co-operating sulphur bacteria, housing them in their guts (in the worms) or gills (in the molluscs) and deriving food from them. It was a seemingly exceptional type of co-operative association between animals and sulphur bacteria. However, submarine springs prove to be quite common even in shallow seas, though they are usually cool, and other sites exist where sulphide-bearing water, not necessarily volcanic in origin, seeps into the sea. Many of the mussels and clams which live in the neighbourhood of these springs and seepages prove to depend on similar co-habitants.

Finally, the gas methane is also common in submarine spring and seepage waters, and bacteria exist which gain energy from oxidising this to carbon dioxide. In the last few years, evidence has accumulated that these types of bacteria, too, can flourish in the gills of marine molluscs and aid their nutrition. Marine molluscs seem quite good at getting exotic bacteria to work for them.

Living extremely close together

Some associations between living creatures have become so intimate that the actual tissues of the partners intertwine. Again, microbes provide the classic example. Lichens, the greenish growth that occurs on rocks, tree trunks and walls, or as splodges of greenish material that grow on trees, are actually combinations of a fungus and another microbe, either a cyanobacterium or a primitive alga (just occasionally both). The non-fungal partner is green – it can photosynthesize – and the two form a structure which resembles an organism, with the green cells on the outside. This is logical, because those cells need sunlight. In some lichens, the cyanobacterial partner can also fix nitrogen, enabling the lichens to colonise remarkably infertile places such as bare rocks and the roofs of houses.

Even more intimate are some associations which are found within protozoa. There is a 'textbook' protozoon called *Paramecium*, a barely visible single cell, rather slipper-shaped, covered with short hairs (called cilia) which enable it to move about. It has a primitive sort of mouth, with which it ingests bacteria as food. It lives in ordinary pond water, and some strains have bacterium-like bodies within their protoplasm which actually provide nutrients for their host. The bacteria seem to have become almost part of the organism, and they cannot be cultured away from it. At least one species of *Paramecium* is often green, because it allows itself to be host to live micro-algae within the cell, which repay it with carbohydrate. Ciliate protozoa which inhabit air-free places, such as rumens or polluted aquatic sediments, almost always have bacteria-like bodies within the cell. Many of these are simply food, bacteria which the organism has recently ingested, but some are permanent parts of the cell. In some of these protozoa the bodies are definitely resident bacteria, because they can be cultured outside their host, and they prove to be methane-producing bacteria. Bodies that have not been cultured have been shown also to be methane producers. They live and multiply along with, but inside, their host cell, having effectively become part of it: a most intimate partnership.

Yet perhaps the most remarkable intimacy occurs within you, dear Reader. In common with all higher organisms, every cell in your body contains tiny structures called mitochondria (singular: mitochondrion). These are the cell's power generators: compartments within which energy is generated, while oxygen is consumed, enabling the cell to grow, to keep itself in good repair, and to do whatever it has to do. All multicellular plants and animals possess mitochondria. There is now good evidence that mitochondria are the evolutionary vestiges of bacteria which, aeons ago, lived in intimate partnership within the cells of some distant ancestor of multicellular creatures. Mitochondria have even retained a few of their own genes over the aeons: mitochondrial DNA duplicates itself independently of, but in synchrony with, the main genetic material, the DNA, in the nuclei of higher organisms' cells, and it has several of the distinctive features of bacterial DNA.

Structures such as mitochondria with an identifiable function within the cells of higher organisms are called organelles. Plant cells have mitochondria, of course, but they also have another organelle, called the chloroplast. This is a different kind of power-house: the plant's solar panel. Within the chloroplast reside the molecules responsible for photosynthesis: the green, light-trapping pigment called chlorophyll, protective pigments called carotenes, and the host of ancillary enzymes needed to trap carbon dioxide and use energy from light to build it up into starch. Chloroplasts make plants what they are. And recent evidence has shown clearly that these organelles, too, are derived from bacteria; this time some kind of cyanobacterium, which formed an intimate association aeons ago with the earliest ancestor of the green plants. Today we know not only that chloroplasts have retained DNA that resembles cyanobacterial DNA, but that this DNA tells them how to synthesise many of their own proteins, and that the structures with which they do this, called ribosomes, are structurally and chemically closely related to the ribosomes of cyanobacteria.

The evidence for the bacterial origin of mitochondria and chloroplasts, over evolutionary time, is convincing. Some other structures, notably certain tubules common to all cells of higher organisms, may have had a like origin. Biologists have long regarded bacteria as present-day representatives of the most primitive forms of life, of the blobs of living matter which first inherited the earth some 3½ billion years ago, so it comes as no surprise to find evidence of that ancestry in animals and plants today. What is unexpected, to me at least, is to learn that our pedigree has not been linear: we have picked up other bacteria on the way and today we are all the collective progeny of microbial associations which, long, long ago, became utterly inter-dependent. That is why, earlier in this chapter, I referred to cyanobacteria as being ancestors of today's plants 'in a convoluted way': they were part of a multiple ancestry. Animals and plants, we are all examples of the accretory, co-operative side of evolution.

What sort of bacterium was our primary ancestor? Microbial evolutionists believe that two lines of descent, into so-called archaebacteria and more ordinary bacteria (called eubacteria), diverged from a common ancestor long before even the most primitive of higher organisms appeared. Such evidence as there is suggests that it was a species of archaebacteria which, absorbing a co-operative eubacterium into its cell, initiated evolution towards protozoa and animals; later a sub-class picked up a cyano-bacterium and became the progenitor of the plant kingdom. (How long ago? I know of no convincing evidence concerning when the two bacterial groups diverged, but fossils of creatures resembling true primitive plants (algae) have been seen in sediments known to be over 2 billion years old.)

Archaebacteria still exist, and their representatives are bio-chemically and genetically very different from both eubacteria and from higher organisms. They include the methane-producing bacteria, some very thermophilic (heat-requiring) sulphur bacteria, and bacteria which need strongly saline environments: nowadays archaebacteria inhabit only hot springs, salt pans and brines, and air-free zones rich in decaying organic matter. (They have several exotic biological features, and examples of archaebacteria appeared often in the earlier chapters of this book.) Their sophisticated and complex descendants have shoved most of them to one side, confining them to the most extreme and exacting of this planet's habitats. The exceptions are the methane producers who, in rumens, in animal guts, in molluscan gills, and within certain protozoa, have craftily re-joined their very distant cousins – from the inside.

16
Immortality and the Big Sleep

Intimations of immortality

That Old Reaper, Death, is going to get all of us in time – you, and me, and every living thing you see around you. Large tortoises, such as you may occasionally see in zoos, may last a couple of hundred years; some trees may last several hundreds of years, even thousands if they are giant Californian Redwoods. But they all die in the end. So it is irritating, to say the least, that some of the things you don't see around you, because they are invisibly small, seem not to. Most bacteria seem to be, at least in principle, immortal.

The study of the mortality of bacteria has not attracted much research during the past half-century. Perhaps I should qualify that remark: of course, an enormous amount of attention has been given to ways of killing bacteria, and some of this research included studies on precisely how a given agent – an antibiotic, a disinfectant, ultraviolet radiation, for example – exerted its lethal action. But the study of if, how, and when bacteria die when subject to no overt stress has been rather neglected, even though it raises important theoretical and philosophical questions, as well as being highly relevant to the survival of microbes in the real world.

Consider, for example, the bacterium called *Klebsiella pneumoniae*. Despite its alarming name, it is rather ordinary; it can cause a rare disease of the lungs, more by accident than design, but it normally lives in soil or in mammalian guts, doing nobody any harm. Individual klebsiellae are nondescript microscopic rods, and there is nothing especially odd about their biochemistry except that many strains can fix atmospheric nitrogen. They grow and multiply best in air, though they can manage without it; they consume a variety of organic compounds and they need no vitamins or supplements beyond the usual minerals: potassium, sodium, magnesium, a spot of iron and so on. In a healthy, multiplying culture they all look much the same – they do not form spores or oddly-shaped variants – and the only obvious feature is that some rods are longer than the others, the shorter ones being in the majority. There is a very simple reason for this: the rods grow until they reach a certain size, when they simply divide across the middle into two short ones, so there are twice as many bacteria that have just divided as there are bacteria about to divide. When division takes place, there is no way of telling which of the two short rods is the mother and which is the daughter. Indeed, such evidence as is available, which in fact applies to *Klebsiella's* cousin, *Escherichia coli*, tells us that no mother–daughter relationship exists: both progeny are of equal age.

Put another way, the question is: are the products of a bacterial cell division, which look alike and behave similarly, mother and daughter? Or twins? In 1958 two US scientists, Mathew Meselson and Frank Stahl, answered that question with an experiment which has become a classic of molecular biology.

The essence of their experiment is this. They grew *E. coli* in conditions in which their sole source of nitrogenous nutrient was an ammonium salt. But it was a special (and expensive) kind of ammonium: its nitrogen atoms were all heavy. (I explained about the heavy isotope of nitrogen in Chapter 11; to recap, it is a scarce form of the element nitrogen which has appreciably heavier atoms than the usual, but otherwise its chemistry is the same). This meant that the bacteria had to make all the nitrogenous components of their cells from heavy nitrogen. Among those components, which were then entirely labelled with heavy nitrogen, was their genetic material: their DNA.

They then allowed some of these 'heavy' bacteria to grow for a generation with a normal nitrogen source.

The crucial question then was, what would happen to the labelled DNA? As each cell multiplied a new lot of DNA would have to be made, copied from the heavy DNA but using normal light-weight nitrogen. DNA is actually a tangle of two long molecules and, during copying, the two strands of this tangle become briefly separated and are copied independently, though simultaneously. Then they tangle up again as two lots of DNA, one for each new cell. So, would all the heavy DNA end up in one of the two new cells? Or would it become divided equally between the two progeny?

DNA with nothing but heavy nitrogen is more dense than ordinary DNA, so Meselson and Stahl exploited this difference. They collected the first and the second generation of cells and extracted their DNA, as well as getting some DNA from cells which had never been exposed to heavy nitrogen. Then they measured the densities of all three. The way they did this involved measuring the buoyancy of the DNA in a strong gravitational field generated in an ultra-centrifuge; precise practical details do not matter here, but an important point is that all three DNAs could be compared in the same centrifuge. The upshot was that the DNA of Meselson and Stahl's second generation of *E. coli* had a density just half-way between the the densities of normal and labelled DNAs. Therefore it must have been a hybrid of labelled and ordinary DNA. It follows that the original labelled DNA must have divided itself equally between the two progeny

of the dividing cells; if it had not, the progeny would have yielded
two kinds of DNA: a *mixture* of ordinary and labelled DNA.

Meselson and Stahl had thus shown elegantly that the heavy
DNA of the parents in the first generation distributed itself evenly
between their two progeny. Since DNA codes for all other cell
constituents, it is overwhelmingly probable that these distributed
themselves in the same way. *Ergo*, there is neither mother nor
daughter; the mother vanished at the moment of division and the
progeny were twins of identical youth.

In anthropomorphic terms, then, when bacteria such as *E. coli*
and, by analogy, *Klebsiella*, multiply, the longer individual vanishes
at the instant of division and two short individuals of equal youth
replace it.

Does the parent cell grow old and die? Of course not. So, do
klebsiellae have any equivalent of senescence and death, analogous to these processes in, for example, ourselves?

Out in the real world, as distinct from in the laboratory, klebsiellae live a life of feast-and-famine, but mostly famine. In soil,
for example, the bacteria are generally starved, or at best strug-
gling with their competitors for the trickle of food which comes
from the slow breakdown of humus – until a surge of nutrient
appears (usually traceable to activities of higher organisms, such
as excretion or decomposition) and then they rejoice briefly in a
surplus. They multiply rapidly until the surplus is consumed,
when they starve again.

During starvation klebsiellae do senesce and die. Some individuals die faster than others. Such death, which seems to be the
nearest they come to a 'natural' death of the kind familiar among
higher organisms, is a response to the mild stress of starvation.
Other mild stresses can cause death – abrupt chilling, a sudden
change in the salinity or acidity of the environment – but starvation is the least drastic of these stresses because it involves the
smallest change in the cells' physical environment. Indeed, in a
biological sense it is the most 'natural' of stresses, because the
microbial world consists almost entirely of bacteria in various
degrees of starvation.

Doing useful experiments on bacterial death is actually rather
tricky. A major reason is philosophical. There is a sort of Biologi-

cal Uncertainty Principle lurking around the study of bacterial death which was put most succinctly by a one-time colleague of mine, Owen Powell: one can only observe bacterial death retrospectively.

What that means is this. To discover how many cells in a bacterial population are dead (if any), the microbiologist has to take a sample of that population, count the cells in it, place it on the surface of a medium in which all living cells can divide, incubate it for a few hours – usually overnight in a Petri dish – and see how many cells have taken advantage of the situation and multiplied to form colonies. Those which have not done so are presumed to be dead. The retrospective element arises when one asks, were they dead when the sample was taken? Or did they die afterwards? Or again, would they have survived and divided if the microbiologist had chosen a different culture medium? All the microbiologist can say with certainty at the end of the test is that a certain proportion of the population was probably dead when the sample was taken.

In other words, there is no reliable way of saying 'that cell has just died'; one can only say 'that cell was probably dead several hours ago'.

In practical work, such as when a food microbiologist is checking for contamination, or a medical bacteriologist is confirming sterility, this uncertainty does not matter: their concern will often be whether any live microbes are present at all, and if numbers are needed, a two-fold error is better than nothing. But such uncertainty can be a discouraging nuisance in pure research, where numbers have to be as reliable as possible. Short cuts to assessing viability have been proposed, based on properties such as how the bacteria stain, look, fluoresce, or respond to another mild stress, but they do not work satisfactorily with mildly stressed (e.g. chilled or starved) populations.

Though these short cuts are often used – not always wisely in my view – there is still no really satisfactory escape from assessing viability by culture; tolerating its element of uncertainty. When my colleagues and I studied death of starving klebsiellae in the 1960s we managed to decrease the period of retrospection substantially: we grew population samples on microscope slides and counted corpses and micro-colonies under the microscope after incubation periods of 1½ to 3½ hours. The ratio gave us the proportion of live to dead bacteria in the sample, though not their absolute numbers; and, of course, uncertainty about whether our slide-culture medium was perfect was always present. However, we got somewhere: our findings were statistically valid and reproducible, and some sort of slide-culture is still the method of choice for studies on the physiology of bacterial death.

It transpires that starved bacteria do indeed go through an ageing process before they die. How long they remain able to multiply depends on how well fed they were before they were starved. Klebsiellae which have had plenty of a carbon/energy food, such as glycerol, but were rather short of nitrogenous nutrients, actually do best; they all survive quite a long time compared with cells which had plenty of nitrogen but were limited in their glycerol supply. Most sensitive are those which grew with limited supplies of magnesium. Death rates during starvation are also affected by how fast they multiplied when they were growing; generally speaking, the faster they grew, the slower will they die. This is all fairly logical, because parallel biochemical tests reveal that, during starvation, they actually consume their own cell material (stored carbohydrate, protein and nucleic acid) and the more of these substances they have to spare, the longer they survive. For example, cells limited in nitrogen have stored plenty of spare carbohydrate, and they survive well using this.

When a typical population of bacteria such as *Klebsiella* is starved, the cells' spare carbohydrate tends to be used first, and some protein is also consumed; then ribonucleic acid (RNA) declines, but deoxyribonucleic acid (DNA: the cells' genetic material) declines only marginally. These processes, which overlap to some extent, provide the cell with energy to help it to survive and, scientists have learned during the last decade, they permit it to make special proteins, called 'starvation proteins', which enable it to survive for longer than it otherwise would. These proteins also improve their ability to resist other stresses such as mild heating or salt damage. But ultimately their reserves run out, and the loss of RNA seems to be the critical thing. Ordinary klebsiellae possess quite a lot of expendable RNA, but when they have consumed about half of their regular complement, they start dying – by which I mean that they become unable to form colonies on a slide-culture.

Yet these 'operationally dead' cells still have many of the characteristics of living bacteria: they respire, albeit slowly; they can resist a change in salinity, keep a poisonous dye out, and sustain their internal pools of amino-acids and ATP. They cannot reproduce, but they retain most of the other features of living organisms: they are like creatures which we would call senescent if they were higher organisms. They appear to be inescapably committed to death.

Thirty years ago, when I and my colleagues first discovered ageing klebsiellae in that state, we thought they were senescent in that sense. But in the early 1990s Douglas Kell and his colleagues, at the University of Wales at Aberystwyth, used new techniques to study a different microbe starved to a comparable state of senescence. They showed that, in special conditions, such cells could be resuscitated. They were not senile but dormant! If other bacteria are shown to behave similarly, dormancy may prove to be an important part of the life of bacteria in the wild.

There is an experimental gimmick whereby microbiologists can actually grow populations of klebsiellae in which many of the cells are in the state we called senescent – with some actually dead. It is called a 'slow continuous culture', but to explain that I must first describe an ordinary continuous culture.

Imagine a culture of klebsiellae in a flask, well provided with air, in a weak solution of glycerol with mineral nutrients: sources of nitrogen, phosphorus, magnesium and so on. It is fully grown, having consumed all the glycerol, but there are plenty of mineral nutrients left. The bacteria, having run out of glycerol, will begin to starve, and in due course some would die. Imagine, however, that the microbiologist admits a little fresh medium, and therefore some glycerol, steadily into the culture and at the same time arranges for an equivalent amount of the culture to leak away steadily – through a tube half-way up the side of the flask, for example. The bacteria will now have a steady, if limited, supply of glycerol and will multiply again. But they will multiply as fast, and no faster, than the scientist allows fresh medium into the vessel. The culture has become continuous: growing steadily, but only as fast as the microbiologist permits. Such continuous cultures are tremendously useful for the study of microbes because the operator can dictate the rate of multiplication, the density of the population (which would be in proportion to the concentration of glycerol in my imaginary example), and the nature of the nutrient which limits the population density (it need not be glycerol; nitrogen, magnesium or phosphorus sources, and even air, can be limiting factors).

Consider, now, a population, limited by glycerol, which is obliged by the researcher to grow more and more slowly, until the imposed growth rate is slower than the death rate of glycerol-limited cells. Then a proportion of the progeny of each cell division dies, leaving some of the in-coming molecules of glycerol un-consumed, which others can use to multiply.

Mathematical modellers have a jolly time with the dynamics of such partly-dead populations – the algebra is not difficult – but I shall leave them to it. A pertinent fact is that, with glycerol-limited *K. pneumoniae* growing at 37° Celsius, a continuous culture growing so slowly that its volume is replaced about every 6 days (which means that the bacterial population in it has had to take about 6 days to double) will be only about 50% viable. Half of them will be in the senescent state, or dead.

Well, in nature (in soil, in a pond, in the treatment plant of a sewage works, for example) the microbial world is rather like a very slow continuous culture, except occasionally when there is a flush of nutrient and they all feast. If all bacteria were klebsiellae, it would follow that most of the bacteria in the biosphere would be senescent, if not dead. But the microbial world is a huge menagerie of diverse creatures, so it would be ludicrous to generalise from one not very special organism to the whole microbial biosphere, would it not?

Yet, would you believe it, it is most probably true.

For all of this century, microbial ecologists have been bothered by the fact that, when they count the numbers of viable bacteria in samples from either natural or man-made environments, using all kinds of culture media, the numbers are always less, by 10- to 100-fold, than the numbers they can see under a microscope, or the numbers they would expect from the level of microbial activity in the environment from which the sample came. Often dismissed as technical error–attributed to rough handling of the sample or inappropriate culture media – it now seems likely that the explanation is not technical: that the populations, though still active, are very largely senescent, with an uncertain proportion of the cells truly dead. There are some types of soil bacteria, such as the arthrobacters, which are remarkably resistant to starvation. Some groups of bacteria can turn themselves into stress-resistant forms called spores, which are wholly immune to starvation (I shall write more of them later), but it is clear that most of the vegetative bacteria which can be seen in soil, sludges and sediments are unable to multiply in the laboratory, though many are nevertheless physiologically active. Whether they are dying or, like Kell's organisms, just dormant and could be resuscitated, is not yet known.

When bacteria finally die, however, the constituents of their cells degrade and, though some material remains within the cell envelope, all sorts of nutrients leak out, to be consumed by surviving neighbours. So in a starving culture one gets a phenomenon known to microbiologists as cryptic growth: a sort of passive cannibalism. For example, when about 50 glycerol-limited klebsiellae have finally died, enough nutrients have been released into the environment from their cells for one of the survivors to grow to maturity and divide. With cells of other nutritional types, as few as five need die to support the doubling of a sixth. So, when a population of klebsiellae forms a colony, in nature or in the laboratory, those crowded up into the air by their neighbours become starved: some die, some become senescent, some multiply by cryptic growth. The same happens in an old fluid culture, or a fermenting brew, in which multiplication has ceased because some nutrient was all consumed, and also in those slow continuous cultures I wrote about. Cryptic growth can muddle long-term experiments of unheeding microbiologists and microbial geneticists; perhaps more important, it undoubtedly plays a part in ensuring the survival of bacterial species during their long periods of starvation in the real world.

How far is death in klebsiellae typical of death in other

microbes? The principle that microbes do not die unless a stress kills them, but split into two offspring of equal youth, probably applies to most bacteria. But there are some microbes that do not split: they bud. Brewers' and bakers' yeasts, which are fungi, reproduce by budding, so that the 'daughters' budded off are younger than the 'mother' which produced the bud. Karel Beran and his colleagues in Czechoslovakia showed in the 1960s that buds leave visible scars, where a new bud apparently cannot form; after producing some 20 buds, a cell was senile: it was covered in scars and so could no longer reproduce. Whether senile cells then died was not clear. Budding bacteria exist as well: the photosynthetic bacterium *Rhodopseudomonas palustris* seems to bud off daughters without restriction, but according to Roger Whittenbury and Crawford Dow of the University of Warwick, its distant relative *Rhodomicrobium vannielli* also buds off daughters but becomes senile after producing only four. Bacteria also exist which grow on stalks, from which they bud off 'daughters' indefinitely. These microbes all show mother–daughter relationships in their reproduction and microbiologists have usually assumed that the senile mothers die. But there is no firm evidence for this: these may be the only potentially immortal beings on this planet, which hang on to life until a stress, self-generated or haphazard, kills them.

Such thoughts provoke reflections, rather different from those of the poet Wordsworth, on immortality. As the great German biologist August Weissman realised as long ago as 1885, the only part of higher organisms such as ourselves that has some kind of immortality is what he called our germ-plasm: the chromosomal material involved in reproduction. The rest, whether it be our mortal selves or a millennium-old tree, is a temporary offshoot of the germ-plasm, which Weissman called the 'soma'. Klebsiellae and such bacteria are different in that their germ-plasm (now recognised as DNA) and soma are within the same cell wall; but is the linear continuity that their (and our) DNA and associated materials enjoy really immortality? Surely those senile mother yeasts and rhodobacteria, which seem to have no reason ever to die, are immortal in a more individual sense? They still have a soma and a germ-plasm, like us; why do we have to die just because evolution has separated our somatic and germ cells? The answer is obvious: because if we and other organisms did not die there would not have been any evolution and we should not be here to ask. But wait a moment – why then has there been evolution at all?

I said the study of microbial ways of death raised important theoretical and philosophical questions. I did not say it answered them.

Let me therefore look at another aspect of death and longevity in bacteria.

Sleeping beauties

Compared with the majority of higher organisms, bacteria do not live for long. Either they divide, or else they become senescent and die. One or the other of these events happens in a matter of hours, days or weeks. There may be bacteria adapted to slow growth in cold environments which can last for a few months before they have to divide or die – we do not really know. Yet despite the short life spans of microbes such as klebsiellae, reports of bacteria of immense age appear in literature of microbiology at intervals.

A surprising one appeared in 1991. The skeleton of a well-preserved mastodon, some 11,000 years old, was found in Ohio in 1989. Associated with the rib cage was a distinct mass of plant material, which seemed to be the remains of its last meal, and from that mass Gerald Goldstein of Ohio Weslyan University isolated a type of bacterium called *Enterobacterium cloacae*. This is a common intestinal bacterium, and it was absent from the surrounding peat. Could they represent the inhabitants of the mastodon's gut when it was alive? It seems very likely, and the question at once arose whether Goldstein's colonies of *E. cloacae* came from individuals which had miraculously survived for the last 11,000 years, or whether they represented microbial families which had persisted for that long in a self-perpetuating microcosm stemming from the decayed gut.

Such tantalising reports are not new. Even in the 1920s, micro-biologists had isolated bacteria from environments such as ancient rocks, and even meteorites, which seemed to have been undisturbed, and unsuitable for microbial multiplication, for centuries and even millennia. For instance, Charles Lipman of the University of California at Berkeley, found 10^8 bacteria per gramme in the interior of a dry bricks from a room in San Luis Obispo mission in California. The room had been used as a prison, then sealed and forgotten for a century. Since the brick was dry and undisturbed, the implication was that the individual cells had survived that long. More startling were Lipman's reports of bacteria in the brickwork of Peruvian pyramids some 4800 years old and, even more amazing, in coal, which would be around 300 million years old.

The microbiologists responsible for such claims have generally been careful. They sterilised the surfaces of their brick, coal, rock or other samples, broke them aseptically, and tested only material from the interior; but nevertheless the implications of their findings have not been widely accepted. The US *Journal of Bacteriology* for 1937 was host to a gentlemanly dispute, a model for the present day, between Lipman and two microbiologists from Washington State University, Pullman, over Lipman's claims regarding coal. His critics argued that recent permeation of extraneous bacteria into the samples would explain his findings. His other claims were not overtly criticised, but there may have been comparable circumstantial explanations. For example, the San Luis Obispo bricks, which contained organic matter, might well have been subject to intermittent periods of sufficient dampness for microcosms of successive bacterial generations to have grown and died *in situ* almost indefinitely.

In other cases, even the precautions taken by experienced microbiologists might not have been sufficient to exclude extraneous bacteria at every stage of the operations. Microbes can multiply on imperceptible traces of organic matter; can survive, even multiply, in minute films of water on particles; and are carried about in aerosols. They penetrate the micropores of geological formations if these are damp, carried not only by capillary flow but also by swimming: from a study of microbes such as *Desulfovibrio* in oil wells, Claude ZoBell of the Scripps Institution of Oceanography in La Jolla, California, calculated that motile bacteria could travel at 2 to 40 feet per year in a deep oil formation. As microbiologists became aware of the versatility and ubiquity of bacteria, they tended to dismiss the reports of such scientists as Lipman as instances of contamination by contemporary microbes. Yet some remained uneasy, especially about such paradoxes as the presence of live bacteria in Arctic permafrost, admittedly not in very large numbers but at sites which may have been frozen for several centuries; some, bafflingly, are thermophilic: only able to grow and multiply at temperatures above 30 to 35° Celsius and doing best at about 65°.

Why were – and are – microbiologists so sceptical? For the reason I have already outlined: ordinary vegetative bacteria are killed very easily by quite minor stresses. That vegetative bacteria should survive in conditions unsuitable for multiplication for more than a few weeks, perhaps months in special cases, is most implausible.

However, there is a way out of the dilemma. Bacteria can go into states of suspended animation, in which they show no, or extremely little, metabolic activity, although they remain capable of resuming their lives when circumstances change. They have three ways of doing this.

One way is by spore formation. A few types of bacteria can sense incipient stress and, in response, undergo a sequence of developmental changes whereby the vegetative cell transforms itself into a relatively dry dormant body covered with a tough coat. This object is the spore; it is spherical or ovoid no matter what the original shape of the microbe was, and within it all metabolic activity seems to be at a standstill. The ability to form spores is not all that common among bacteria: it is rare outside the genera called *Bacillus* and *Clostridium*, and a group called Actinomycetes. But when they are formed, bacterial spores are very resistant to freezing, drying, chemicals and heat; some even survive brief pressure-cooking, and spores of the most heat-resistant type, *Bacillus stearothermophilus*, are used routinely to check the efficacy of hospital autoclaves.

A few other bacteria form comparable bodies which, although

more resistant to maltreatment than the vegetative cells, rarely approach spores in toughness. Those of Cyanobacteria are called akinetes, those of *Azotobacter* are called cysts.

The other two means of becoming dormant are passive processes, outside the microbes' control. Drying or freezing kills the majority of vegetative bacteria but, in a variety of special environments, some or all of the cells may retain ability to resume their lives if and when they are rehydrated or thawed as the case may be. Naturally, in the dry or frozen states, they show no metabolic activity.

Bacteria can remain viable in the dry state if the environment contains protective agents which dry along with them. Sugars (especially when mixed with protein or another polymer) are good protectants. A mixture of glucose and sterilised serum is especially versatile in protecting a wide variety of bacterial genera and is used by culture collections to preserve their stocks in a dried (generally freeze-dried) condition. (The mixture is called *mist. desiccans*; an archaic name with appropriate alchemical overtones because the reason why it works so well is still something of a mystery.)

When a population of bacteria dries out *without* a protectant, many of the cells break open and release their internal contents. Among these contents are proteins, gums and sugars, all of which are protective. If the population is suitably dense, so that significant amounts of protectants are released, material released from the majority which died first can protect a few of their surviving fellows.

Comparable considerations apply to death from freezing, of which I wrote more in Chapter 3. Protective substances such as glycerol are well known and widely used; they are called cryoprotectants. Bacteria frozen without such chemicals leak internal contents, among which are many substances that are cryoprotective.

Bacterial populations of sufficient density thus protect some of their members from freezing or drying at the expense of others. In nature, bacterial populations may well protect each other – I do not know of any study of this question – but the natural environment is certainly fraught with agents which have protective actions, such as carbohydrates, proteins, and soluble degradation products of these and other large molecules. Soil, for example, is a good protectant against both freezing and drying. The early soil microbiologists – and some wise contemporary ones – used this knowledge to store specimens of their isolates. They would add a few drops of a culture to sterilised, dry soil, allow it to soak in and store it in a cupboard. Often the bacteria survived there for decades. Bodily exudates and excretions from animals, decomposing organic matter and so on, yield protectants. Reflect, if you will, on what happens as nasal mucosa on your pocket handkerchief dries in your pocket (happily you and your own nasal micrococci usually get on well with each other).

Drying out, on the one hand, or becoming frozen, on the other, in the presence of a natural protectant are the only ways in which vegetative bacteria – those which cannot, or did not, form resistant bodies – become dormant. Goldstein's *Enterobacterium cloacae* cells from the mastodon are closely related to *Klebsiella* and *Escherichia coli*, and none of them can form resistant bodies. How long do such dormant bacteria actually remain viable? The question has not been studied for long enough for a clear answer to be given. As far as the frozen state is concerned, the general pattern in the laboratory seems to be that, when a population of vegetative bacteria is frozen in the presence of a cryoprotectant, almost all will survive at first but then the population dies slowly: very slowly indeed with a good cryoprotectant such as glycerol, but quite rapidly with a less good one such as *i*-erythritol. Lower temperatures favour longer survival. Dried or freeze-dried populations seem to lose viability slowly, too. In both cases there is evidence that, with some species, a few of the original bacteria survive in the frozen or dry state for more than four decades, but the rates at which viabilities decline suggest that a limit is being approached. As an informed guess, a few out of an initial billion might last a century, give or take a couple of decades.

Therefore, reflecting on Goldstein's *E. cloacae* in the gut of the mastodon, even if cold or dehydration had helped to prolong their lives, it is difficult to believe that they could have lasted 11,000 years; the idea that the decaying gut provided a long-lasting microcosm for descendants of the mastodon's *E. cloacae* is more compelling.

With spores, the prospects for longevity are greater. In 1962 Peter Sneath, then working at the National Institute for Medical Research at Mill Hill, London, looked into the matter. He knew that a suspension of spores of the anthrax bacillus prepared by Louis Pasteur in 1888 was still viable in 1956, and that a 118-year-old can of meat had been found to contain spores of a thermophilic microbe. He also mentioned that 'coliform bacilli' had recently been found in frozen faeces deposited by Scott's ponies on his 1911 Arctic expedition. However, he himself had examined an approximately 3500-year-old coprolite from Mexico and found it sterile. These were isolated examples – was a systematic approach to the longevity of dormant bacteria feasible? In a felicitous piece of lateral thinking, he realised that, in the Herbarium at Kew Gardens in London, there existed a collection of dried and individually packaged plant specimens which had been added to regularly since 1640. The soil associated with these should contain the spores of bacilli of a wide range of ages. They did. Sneath found viable bacillus spores, in decreasing numbers, in samples up to 320 years old. Plotting their specific numbers, he constructed a survival curve and, from published data on numbers of bacilli in ordinary soil, he predicted that a ton of dried soil would retain a few viable *Bacillus* spores even after 1000 years.

This experiment provided a frame of reference for other studies. It had only one tiny flaw: the soils on the plant specimens would have had different populations of bacilli when fresh. However, the differences would have been statistically trivial.

Not all reports have been as rigid as Sneath's. In 1963 Heinz Dombrowski of the Justin Liebig University, Germany, claimed to have found viable *Bacillus circulans* in samples of salt from Devonian, Silurian and Permian deposits, originating in Germany and North America. This would have implied ages in excess of 600 million years; his findings were immediately disputed, because, though this microbe could hardly have multiplied in crystalline salt, its spores are common in air.

Another surprising report came in 1966 from Yusef Abd-El-

Malek and his colleague Ishac, at the University of Cairo. They found viable clostridia and azotobacters in the interior of a mud-brick from the Great Temple of Amon at Karnak, Luxor, Egypt. The brick came from a height of 15 metres and was protected from rain. It was at least 2400 years old yet the straw which had been used in its manufacture had still not rotted. Azotobacters were few – 11 to 17 *Azotobacter chroococcum* per gramme of brick dust – but for such longevity to have been found at all was hard to credit, because azotobacters do not form spores but only the much more fragile cysts. They are also notoriously sensitive to drying, though they actually survive that stress best in soil. However, the soil of the Nile Delta is notable among microbiologists for the highest recorded densities of azotobacters in the world, so it is difficult not to conclude that modern bacteria, on dust from the delta perhaps, had crept into the experiment.

In 1976–7 Tom Cross and his colleagues at the University of Bradford found viable spores of a group of thermophiles called *Thermoactinomyces* in occupational debris beneath the Roman fort of Vindolanda, five miles south of Hadrian's Wall, near Carlisle. The deposits date from around AD 90. They also found spores of *Thermoactinomyces* in lake sediments, in East Anglia and the Lake District, of dates ranging from about 600 BC to about AD 50. In 1978–79, Nancy Parduhn and John Watterson of the US Geological Survey found spores of *Thermoactinomyces vulgaris* in stratified sediments in Elk Lake, Minnesota. They carbon-dated their samples; spores were there in substantial numbers (some 1000 per gramme) in a stratum about 5150 years old, with lower concentrations in younger and older strata. The oldest stratum approached 7200 years old.

There are three possible ways by which modern spores might reach ancient sediments or soil and so delude microbiologists. They themselves, or viable fragments of their vegetative forms which had grown elsewhere, could be carried in by permeating water. They might also be transported by worms and other fauna; and finally, they might have grown there. However, the workers at Elk Lake, in particular, excluded transport, arguing that sediment fauna would have disturbed the strata, which were clearly undisturbed, and further that a peak in their distribution at 5150 years old with a trough on either side could not arise from physical permeation. Both groups of workers excluded growth *in situ* on the grounds that *Thermoactinomyces*, being a thermophile, is unable to grow in cold soil or sediment – yet a devil's advocate might contend that, though thermophiles do not multiply below 35° Celsius in laboratory tests lasting weeks or even months, who really knows that they do not metabolise and even multiply, albeit imperceptibly slowly, over several centuries out there in the unchanging cold?

An earlier report by J. W. Bartholemew and George Paik of the University of Southern California, Los Angeles, in fact took care of this last point. In 1965 they were studying the floor of the Pacific Ocean off Mexico and discovered viable spores of *Bacillus stearothermophilus* in cores of marine sediments at a constant 4°, cores which carbon dating had established ranged from 6000 to 8000 years old. Their incredulity was reflected in the almost hesitant tone of their publication. As well as being a thermophile, *B. stearothermophilus* is not a marine organism, so multiplication *in situ* can be excluded because the sediments were too salty as well as too cold. But in this case transport could not be rigidly excluded.

One can niggle about details of the work on sediments, but taken as a whole the conclusion is compelling that the spores settled there as the sediments were forming, delivered by aerial deposition or in terrestrial run-off water in the marine case, and have survived there ever since: for up to 8000 years.

If spores can live so long, why do they die at all?

Sneath reflected on this problem and noticed that the decay curve for his bacilli from Kew was not unlike the decay curve of a mixture of radioactive isotopes of different half-lives. All living material is subject to natural ionising radiation, from such sources as cosmic rays and radio-potassium: might the slow death of spores be assignable to background radiation? At Oxford a few years ago Howard Gest and Joel Mandelstam decided to look into the matter. Ionising radiation kills most of the bacteria in a population, but some survivors prove to be genetic mutants. If spore suspensions die because they are affected by natural radio-activity, then old spore suspensions ought to contain more mutants than fresh ones. Gest and Mandelstam had a 16-year-old suspension of spores of *B. subtilis* so they compared the number of mutants in that population with the number in a similar but fresh spore suspension. They chose to look for a class of mutants called auxotrophs, in which the microbe develops the need for a growth factor in order to multiply, because they are easy to detect and score in the laboratory. The two populations proved to be essentially identical: about five spores per thousand were auxotrophic in each case. This is almost spot-on the 'spontaneous' mutation frequency which microbial geneticists obtain with their workhorse, *Escherichia coli*, when it is *not* treated with radiation or otherwise mutagenised; that the two populations were the same implies that background radiation has no detectable influence on the viability of *Bacillus* spores.

So there seems to be no good reason why they should die at all. Perhaps they do not; perhaps they are immortal. Peter Sneath's organisms at Kew, for example, may, now and again over the centuries, have become sufficiently damp and warm, on humid summer's days perhaps, to initiate germination – only to die of starvation. Spores deep in soil or in sediments are generally screened from such fluctuations, except over geological time, when glaciations, warmings or tectonic activity might have relevant effects.

There may be much older spores out there, waiting for energetic microbiologists to revive them.

17

Self-adjustment

The genetic blueprint

The 1960s and 1970s were glorious decades for biology. The nature of genes, the chemical packages of information with which living things transmit their heritable characters to their offspring, had become clear; the code in which that information is stored had been cracked, and the means whereby it is transcribed and eventually translated into a living organism was well on the way to being understood. Almost all these marvels had been learned from the study of a single species of bacteria called *Escherichia coli*, a normal inhabitant of the guts of mammals (including ourselves). The information seemed, at the time, to apply to all living things – at least in outline. A laboratory joke of the period was to claim that what is true of *E. coli* is true of the elephant. I doubt whether anybody ever really believed that, but the research of those years was certainly revealing that, in addition to the genetic machinery, the fundamental pathways of metabolism, such as the manufacture of proteins, and the utilisation of carbohydrates, are indeed very similar in the cells of almost all living things. It was – and still is – fair to say that *E. coli* and elephants have a lot in common in the way their cells work.

Yet oddly enough, it is at the genetic level that some of the most striking molecular differences between *E. coli* – representing bacteria in general – and elephants – standing in for higher organisms – have turned up. Although bacteria and higher organisms use essentially the same genetic code, they differ substantially in the ways in which they package that information in their genes, and in the way in which they translate that information into synthesis of cell material. They differ, too, in that bacteria have a remarkable number of ways in which they can transfer their genetic information, not only between species but, unique among living things, between one genus and another.

A word about genes. I mentioned in Chapter 2 that genes are stretches of a chain-like molecule known as DNA and, for that chapter, the fact that the chains are paired and coiled was important. Here I am more concerned with the units of which each chain is composed, which are called nucleotides. These are quite complicated molecules, They include a kind of sugar called deoxyribose, phosphate (a combination of phosphorus and oxygen), and a molecular structure called a base (comprising carbon, hydrogen, oxygen and nitrogen), linked together as a trio. In living cells, the phosphate group of one nucleotide links up with the sugar of another, whose phosphate is attached to the next, and so on, to form the DNA chain. A chain of about a thousand nucleotides makes up the average gene.

There are four kinds of nucleotide in DNA, made from four kinds of base. The bases prove to be the salient features of the genetic code: they are the letters, so to speak, that spell out the information encoded in the gene. The cell has an alphabet of four letters, and it so happens that all its words are three-letter words: trios of bases carry the information, which is all about what proteins the cell should make. Apart from some punctuation, which I shall come to shortly, this is all the hereditary information a cell needs, because the majority of proteins, being enzymes, are the catalysts which enable the cell to metabolise, grow, move, sense and do all the other things that constitute life.

Proteins are chain-like molecules, too, and they are made up of units called amino-acids, of which twenty occur in ordinary proteins. These, in various combinations, determine what sort of protein it is and what it will do. If one imagines DNA as the tape of an audio-cassette, then the cell has structures within its protoplasm which act rather as a magnetic head does in a cassette player, reading the information along the tape and sending its messages to be processed – not into sound but into instructions telling the protein-making apparatus what amino-acids should be used by the cell and in what order. Successive trios of bases along the gene specify successive amino-acids, which the protein-making machinery sticks together sequentially to build up a protein chain.

I mentioned punctuation. A chain of DNA usually carries thousands of genes, so the 'start' and 'stop' signals must be clear. In *E. coli*, genes are separated by short stretches of non-coding DNA, in which there are special clusters of bases saying 'start here' (or rather, 'start soon after here'). Similarly, at the end of a gene, when trio after trio of bases have coded for all necessary amino-acids, there is a trio saying 'stop here'. One gene specifies one protein.

The elucidation of the genetic code by Dr. Francis Crick and Dr. Sidney Brenner in the 1950s was a major triumph of molecular biology. The permutations and combinations of four bases are more than sufficient to code for 20 amino-acids plus a stop sign. In fact, a triplet code based on four letters could accommodate 64 bits of information (e.g. 63 amino-acids and a stop). In reality, there can be two, three and, in a couple of cases six, different trios signifying the same amino-acid (a little like Elizabethan spelling . . .), and there are three stop signs.

In higher organisms, plants and animals, genes are structurally more complicated than in *E. coli* because they are interrupted internally by non-coding stretches of uncertain function. Moreover, the start and stop signs are different and the DNA is segregated within the cell's nucleus, Finally, they are packaged in a multiplicity of differing pairs of chromosomes whereas bacteria have only one. But the 'alphabet' is the same.

A spot of sex

To return to genes themselves. Genes maintain the stability of a biological species, ensuring continuity between parent and offspring. But variation is the stuff of evolution: for species to adjust to changing environments, or for new species to emerge, genes themselves must change.

The most radical way in which hereditary change occurs is by mutation. When DNA is being copied, as must happen when a cell reproduces, mistakes can occur. Some happen for the simple reason that no copying sequence of that complexity can work wholly without error. Some occur because a chemical which reacts with DNA, or a physical agent damaging to DNA (such as a pulse of short ultraviolet radiation), has fortuitously apppeared. Because genes are double stranded, the cell can correct most errors by reference to the other strand, but some get through, and then the gene is permanently altered. Most often it is inactivated, and the progeny cell dies; sometimes the mutational change is minor or trivial; occasionally mutation produces new and biologically valuable genetic information. Mutation is a rare and haphazard process, yet it is the principal way in which terrestrial life has evolved, and continues to evolve, to ever-increasing complexity.

The common mode of variation among higher organisms, as everyone knows, is through sexual reproduction. Even within well-defined species, accumulated minor mutations cause genetic patterns to differ among families and races. By doubling up their quota of genes, and providing their offspring with one set from each parent, organisms which reproduce sexually ensure regular mixing of the 'gene pool' within a species. Thus, constant, if small, biological change occurs from generation to generation, and natural selection imposes direction on such changes over the long run.

Among higher organisms today, sex, underpinned by minor mutations, is the machinery of evolutionary development. But among bacteria, things are different; they are sexless and, with a very few exceptions, they multiply by fission: a parent splits in two to yield two daughter cells of equal youth. But despite being sexless, they do transfer genetic material from one organism to another. *E. coli*, for example, has three modes of gene transfer. To explain them I must use three technical names.

Transduction

Bacteria, like other living things, are subject to virus infections, some trivial, some lethal. Many bacterial viruses consist largely of DNA themselves, and some of these, after they have infected an *E. coli* cell, combine with bits of their host's DNA and make it part of themselves. As the virus multiplies, that piece of host DNA multiplies along with the virus DNA. In due course the host dies, and releases into the environment a hybrid virus carrying fragments of its DNA. When the hybrid virus attacks and enters a different bacterium, it carries in that DNA. Most of the cells attacked will succumb to the infection, but a few will be resistant, and these will integrate the alien DNA in among their own genes. If the new DNA includes DNA sequences which the resistant host can use, it will do so, and its genetic make-up will thus have been changed. The pieces of DNA transduced by bacterial viruses may amount to one or two whole genes, but they are often smaller fragments which may nevertheless modify genes already present in the recipient.

Transformation

E. coli at a certain stage of their growth cycle can, after certain physical or chemical stresses, take up raw DNA (such as DNA purified in the laboratory) and incorporate them into their own genomes. Transformation can involve quite large pieces of DNA, comprising dozens of genes, such as genetical elements called plasmids (below).

Conjugation

E. coli only conjugate when one of the cells possesses so-called fertility genes and the other does not. The two organisms come together, a tube grows between them, pulling them close, and the fertile strain donates fertility genes to the recipient, which then becomes fertile itself. The fertility genes are always accompanied by genes coding for other properties, sometimes many of them. Conjugation has been held to be a primitive form of sexual congress, the donor cell representing the male and the recipient the female, but it is an odd sort of sex in which the recipient becomes male and both parties can multiply!

Conjugation may involve the transfer of anything from about 30 to over 100 genes. Genes coding for fertility most often reside, not on *E. coli*'s regular chromosome, but on genetic elements called plasmids. These are mini-chromosomes which exist within the cell together with the regular chromosome; in *E. coli*, plasmids range from 3% to about a third of 1% of the size of the chromosome. Most often there is one copy of a given plasmid per chromosome, but with the small ones there may be several copies. Sometimes two or more kinds of plasmid are present in a given strain of *E. coli*, though plasmids seem to belong to groups, some of which are not compatible with others in the same host. Many plasmids do not carry fertility genes, but some of these can be mobilised by plasmids which do have fertility genes and be co-transferred to a recipient strain.

Plasmid DNA may code for a wide variety of physiological properties including resistance to anti-bacterial substances, ability to make an antibiotic, to make a toxin, to metabolise various substances, as well as fertility.

Numerous plasmids have been found in, or introduced into, *E. coli*; the species seems to carry its main genetic information on chromosomes, but to have a bank of supplementary information stored on plasmids, distributed among a variety of *E. coli* strains but never all present in the same one. Put another way, among the world's population of *E. coli*, the gene pool of its chromosomal genetic information is substantially larger than the gene pools of its plasmid-borne information.

This seems a curious way for a species to handle its genetics, but it has a certain logic. The main habitat of *E. coli* is very stable; it is the lower intestine of mammals. (*E. coli* has another habitat, soil and fresh water, but in these it is a transient inhabitant: it reaches such environments in faeces and is usually competed out by indigenous microbes in a few hours or days.) Current thinking is that the chromosome of *E. coli* provides all it needs in its normal primary habitat but, if it encounters a stress – a changed substrate, an antibiotic, a competitor – plasmid-borne information will be available among a minority of the population. Therefore the species, if not the majority of individuals, will survive.

That rationalisation implies that its chromosomal information ought to be stable. But there is good evidence that it is not; there seem to be several ways in which the genes on *E. coli*'s chromosome can become rearranged or modified. In the first place, whole plasmids may, in special circumstances, become integrated into the chromosome, pushing neighbouring genes aside and silencing the one where they insert themselves. Secondly, many of the genes carried by plasmids, such as those specifying resistance to antibiotics such as kanamycin or penicillin, are flanked by special DNA which enables them to transpose from plasmid to chromosome and back, or from one plasmid to another, again silencing a chromosomal gene where they insert themselves. These 'jumping genes' are called transposons. Thirdly, the chromosomes of *E. coli* contain a number of enigmatic lengths of DNA called 'insertion sequences' which code for no known product, but which may move about the chromosome during multiplication, again silencing or modifying genes. Insertion sequences are present in plasmids, too.

Gene-transfer processes and their capacity for gene rearrangement provide bacteria with the potential for substantial and rapid genetic change. But there is yet another consideration. Mutation is a very rare event as far as higher organisms are concerned, but is an important part of the everyday biology of bacteria such as

E. coli. The reason stems from the fact that the bacteria are extremely numerous and have rather a small complement of genes. Let me illustrate the point with a simple calculation concerning the global population of *E. coli* living in the human gut.

Humans on an average discharge about 200 grammes of faeces a day. Human faeces contain a surprisingly constant figure of about 100 million *E. coli* per gramme. Thus a single person produces about 20 billion *E. coli* per day, almost all freshly grown since the last defaecation.

In 1990 there were about 5 billion humans in the world (there are lots more now). The global growth rate of our intestinal *E. coli* was thus about 100 billion billion (10^{20}) cells per day.

In the last couple of decades, molecular biologists have established that *E. coli* possesses enough DNA for about 4000 genes, which seem to be all that it needs to be *E. coli.*

Mutations which do the organism no serious harm (such as those which lead it to acquire resistance to a drug, or to need some fairly common nutrient) occur spontaneously in multiplying *E. coli* at frequencies in the range 1 per 100 thousand to 1 per billion of new progeny. For the sake of argument, say 1 per 10 million.

Obviously, therefore, more than 10 thousand billion *E. coli* genes mutate daily inside humanity. Each mutation signifies an altered gene. Since *E. coli* contains only 4000 or so genes, it follows that *every gene of the gene pool of E. coli* traversing the human intestine mutates at least 2.5 billion times daily.

That calculation makes some crude approximations, especially over the spontaneous mutation rate. But it is not more than an order of magnitude wrong: the message is that the daily mutability of *E. coli*'s collective genetic material is astronomical. Add in the *E. coli* inhabiting the guts of other mammals, and be informed that stresses experienced when *E. coli* leaves the gut increase its mutation rate, and the number of mutations gets even larger. The upshot is that among the terrestrial population of *E. coli*, every possible mutation is occurring, an enormous number of times a day.

Fluid genetics

Do the genetic principles I have described for *E. coli* apply to microbes with less stable habitats?

In the last decade microbiologists have come to realise that, though what is true of *E. coli* may not always be true of the elephant, it is usually true of other bacteria. In particular, the genetic flexibility shown by *E. coli* seems to be common. Transformation was in fact first discovered in pneumococci and largely worked out in the genus *Bacillus*; its discovery in *E. coli* is relatively recent, and it now seems that most bacteria can be transformed by raw DNA if it is in the form of a plasmid. Transduction of genetic information by viruses occurs in many groups of bacteria. And some other means of gene transfer have been discovered, such as cell fusion among bacilli and an enigmatic gene-transfer agent in certain photosynthetic bacteria.

Perhaps most significant, microbiologists have also come to realise that plasmids, once thought to be exceptional, are so common as to be almost the rule among bacteria. Many are 'cryptic', which simply means that their discoverers have no idea what their DNA codes for. Some of them are huge: up to 30% of the size of the chromosome; others may be tiny, with room for only about half a dozen genes. Often there are several different-sized plasmids in a single cell. Not all are cryptic. In a bacterium called *Rhizobium leguminosarum*, which colonises the roots of pea plants, genes for both recognising the right species of host plant and for fixing nitrogen are on plasmids. A plant pathogen called *Agrobacterium* has plasmids, one of which specifies its pathogenic ability. Among the soil and water bacteria called collectively pseudomonads there are plasmids enabling them to metabolise exotic chemicals such as toluene, camphor or oil hydrocarbons, as well as plasmids coding for resistance to medically important drugs. Some carry fertility genes which enable pseudomonads to mate among themselves, and also to pass the plasmids to unrelated – or at least only distantly related – bacteria, including *E. coli*; they have been called, evocatively, 'promiscuous' plasmids. These, carrying their transposons and other genes, are not sensed as alien, and the recipients can usually pass the plasmids on again. In fact, many of the plasmids used in research on *E. coli* were first found in *Pseudomonas* or *Salmonella*; the question in what species a given plasmid first originated can be a difficult one.

How important are these gene transfer processes in the natural environment? In hospitals and in animal husbandry, epidemics of resistance to therapeutic drugs spread by plasmid transfer among pathogenic bacteria is a recognised problem. Recent ecological studies, stimulated by anxieties about the release of genetically engineered bacteria, indicate that gene transfer between species and genera takes place in nature. Especially interesting in this context are some experiments by Brian Spratt and his colleagues at the University of Sussex, which showed that, in a couple of groups of bacteria, strains of naturally penicillin-sensitive species had acquired ability to resist penicillin by 'recruiting' stretches of DNA conferring penicillin resistance from different, naturally resistant, species. The process involved was probably transformation.

It is likely, too, that the chromosomes of all bacteria are as mutable as that of *E. coli*. Their chromosomes range in size from a half to three times that of *E. coli* and, though some yield mutants less readily in the laboratory than does *E. coli*, this is because some are better able than others to repair the mutations. In principle, therefore, the calculation I did for *E. coli* applies to all bacteria. A slow-growing soil bacterium, good at DNA repair, might require weeks, even a couple of months, for its global gene pool to undergo as many mutations as *E. coli* gets through in a day, but even so, the potential of bacteria for rapid mutation is phenomenal.

I hope I have conveyed some impression of the tremendous genetic fluidity in *E. coli* which is brought about by transduction, transformation, plasmid transfer, transpositions, insertions and mutations, and the implications of this for the rest of the bacteria. Because now a note of caution is necessary. About a decade ago certain transatlantic microbiologists were so impressed by such revelations that, in what they called a 'Manifesto for a new bacteriology' (a charming echo of the cultural manifestos beloved of early twentieth-century artists), they advanced the view that the bacterial world is a super-organism: a single planetary entity. But in truth, despite the genetic flexibility of bacteria as a whole, independent, autonomous bacterial species are the norm: the strains and groups which microbiologists call species and genera behave as stable biological hierarchies and possess physiological and genetic equipment enabling them to sustain their specific integrity. If alien DNA enters *E. coli* by transformation or as a virus, for example, *E. coli* can recognise alien features and destroy the unwanted DNA using special enzymes called restriction enzymes, which chop it up into pieces which the cell can degrade and excrete (these enzymes have been isolated and have proved exceptionally useful for genetic engineering; that is, for manipulating DNA in the laboratory). And there are other mechanisms in bacteria for rejecting unwanted DNA.

Evolutionary considerations

A close relative of *E. coli* is *Salmonella*, a bacterium whose primary habitat is the intestines of birds (notoriously chickens and ducks) and which, like *E. coli*, inhabits soil and fresh water transiently. The chromosomes of *E. coli* and *Salmonella* are very similar – about 80% alike – which means that there are long stretches of DNA which are identical in the two organisms. *Salmonella* is genetically as flexible as *E. coli*, undergoing transduction, transformation and conjugation, and the two organisms exchange plasmids readily. However, there is a barrier to the exchange of chromosomal DNA; if, by experimental ingenuity, chromosomal DNA from *E. coli* is introduced into *Salmonella typhimurium*, the DNA repair system of the *Salmonella* recognises that something is wrong and 'tidies up' the alien DNA. In some way *Salmonella* knows it is not *E. coli*, but it would not require many mutations for *Salmonella* actually to become *E. coli*, or the reverse, and, given the global mutation rates of both genomes, it would not take long, given the right selection pressure.

Howard Ochman and Allan Wilson of the University of

California, Berkeley, studying the resemblances between *E. coli* and *Salmonella*, concluded that the two species diverged about 130 million years ago, 'intriguingly' (they wrote) at about the time of the origin of the mammals. They have remained distinct, yet recognisably close relatives, ever since because their habitats have not changed significantly.

That is one reason why microbiologists continue to find organisms which they can identify as *E. coli* when they examine human faeces or environments polluted with sewage. But an important supplementary reason is that the genetic flexibility of this particular microbe is constrained by its primary habitat, the mammalian gut. From the point of view, so to speak, of their bacterial inhabitants, the guts of mammals have changed but little, and very slowly, over many hundreds of millennia. Therefore they have conferred a veneer of stability on their guests because, though many of the genetic fluctuations of which *E. coli* is capable undoubtedly do occur, all the really advantageous variations took place long, long ago and the current variations are either trivial or disadvantageous, so the variants die out of the population. The guts of different varieties of mammal will impose marginally different patterns of genes; no doubt that is why biochemical families of *E. coli* exist, differing in the fine details of their enzymes. More practically, it is probably one reason why the *E. coli* from the human inhabitants of distant lands, where different diets provide it with significantly distinct primary habitats, are apt to give the intrepid traveller diarrhoea.

All this potential for gene exchange and mutation in no way challenges the role of natural selection in bacterial evolution, but it implies that evolutionary change, sometimes in big jumps, can be expected, given appropriate selection pressure, in matters of days, weeks or months, rather than the tens of millennia observed among animals and plants. The late Robert Hedges, of the Royal Postgraduate Medical School at Hammersmith, a pioneer of plasmid research, suggested almost 20 years ago that bacterial evolution has not been linear, as in higher organisms, but rather a patchwork, with organisms drawing from a communal gene pool. As molecular genetics has advanced in the intervening years that view has come to seem increasingly close to the truth. Mechanisms to restrict the intake of alien DNA, and to repair damaged DNA, confer a certain degree of integrity on the genes of bacterial species, but these systems mutate, too. It follows that the world of bacteria at any stage of this planet's history has been conditioned by the state of their habitat. If this has been stable, its microbial population will have been stable; if it has been changing, its bacterial population will have changed along with it.

It also follows that bacteria which live in association with higher organisms, such as *E. coli* in our guts, or the fermentative bacteria which live in the rumens of cattle, or those that inhabit the root nodules of plants, will have evolved along with their hosts, in directions determined by their hosts. Even the evolution of pathogens, which harm their hosts, will have been much influenced by their hosts. In the broad context of the origin of today's flora and fauna, their evolution is not very interesting.

Well, perhaps that is unfair, but to me things get considerably more exciting when one considers free-living bacteria, those which inhabit soil, sea, lakes and so on. Then a different picture emerges.

Present-day speculations on the very early history of the Earth seem to agree that, during the first 2 billion or so of life's 3½ billion years of existence here, tremendous changes took place in the chemistry of the planet's surface and atmosphere. Some of the changes were the result of geological upheavals consequent on the Earth's formation: earthquakes, storms, inundation, volcanic eruptions. But more radical, if more gradual, changes were wrought by living things, and the majority of these changes occurred before the multicellular creatures appeared, when the dominant flora comprised bacteria. It was then that creatures emerged with bizarre habits such as I discussed in the first few chapters of this book. For there was effectively no oxygen, so they were all anaerobes, and hot, sulphurous and salty places were abundant, and there were no plants or animals, though there would have been modest amounts of organic matter formed by spontaneous chemical reactions: cyanides, hydrocarbons and even amino-acids.

It was then, when pristine bacteria were transforming the world, that major steps in the evolution of bacteria took place:

new microbes came into being and colonised new zones of the planet, acquiring ability to reduce carbonates and form methane, to 'breathe' sulphate rocks and form sulphides, to metabolise sulphur and sulphides, to gain energy from iron compounds, or from sunlight. Quite possibly the earliest types were like the thermophiles I wrote of in Chapter 2, organisms which grow only in hot water, and the pristine creatures may have had to be very tolerant of acidity. But gradually these bacteria-like creatures acquired much of the extraordinary chemical versatility familiar to microbiologists today, and in doing so altered the chemical nature of the environments in which they lived.

In return, so to speak, that changing chemistry would have conditioned the directions of bacterial evolution. For example, the earliest photosynthetic bacteria were almost certainly anaerobes: they would have used sulphides or organic compounds in their photosynthetic reduction of carbon dioxide. But in due course bacteria able to conduct the sort of oxygen-producing photosynthesis that plants exloit today would have emerged, probably descended from one or other of the photosynthetic anaerobes. The first kind would have resembled blue-green bacteria, which are common today, called cyanobacteria, and there is indeed evidence for traces of microbes that look quite like them in very ancient rocks.

The crucial importance of the emergence of oxygen-producing photosynthetic bacteria cannot be over-emphasised. Oxygen had long been a transient component of the atmosphere, but an unstable one, because it had quickly been absorbed by chemical reactions with iron and carbon compounds. But oxygen-producing photosynthesis, once it was established, was so efficient that it ultimately caused oxygen to become a permanent component of the atmosphere. This happened no earlier than a billion years ago, though the timing is much debated (the first true plants appeared about half a billion years ago). Air with oxygen represented a drastic transformation of the world; the oxygen would have poisoned most anaerobes, doubtless extinguishing many types, and duly brought about this planet's first population explosion: bacteria that learned to make use of oxygen, who learned to respire as we do today, inherited an Earth rich in

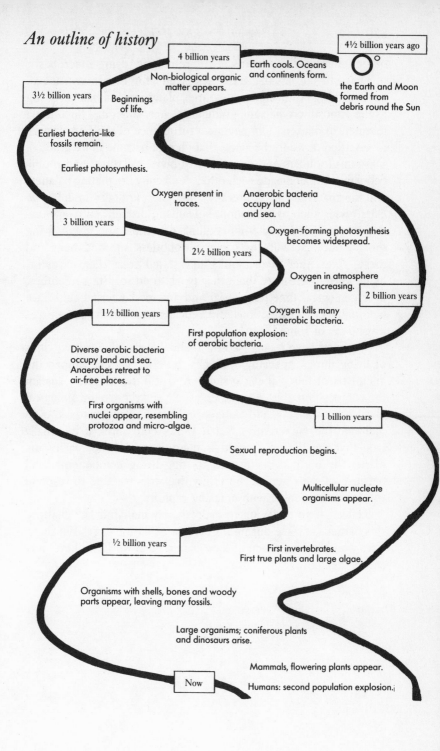

An outline of history

4½ billion years ago

the Earth and Moon formed from debris round the Sun

4 billion years

Earth cools. Oceans and continents form.

Non-biological organic matter appears.

3½ billion years

Beginnings of life.

Earliest bacteria-like fossils remain.

Earliest photosynthesis.

Oxygen present in traces.

Anaerobic bacteria occupy land and sea.

3 billion years

2½ billion years

Oxygen-forming photosynthesis becomes widespread.

Oxygen in atmosphere increasing.

2 billion years

Oxygen kills many anaerobic bacteria.

1½ billion years

First population explosion: of aerobic bacteria.

Diverse aerobic bacteria occupy land and sea. Anaerobes retreat to air-free places.

First organisms with nuclei appear, resembling protozoa and micro-algae.

1 billion years

Sexual reproduction begins.

Multicellular nucleate organisms appear.

½ billion years

First invertebrates.
First true plants and large algae.

Organisms with shells, bones and woody parts appear, leaving many fossils.

Large organisms; coniferous plants and dinosaurs arise.

Mammals, flowering plants appear.

Now

Humans: second population explosion.

organic matter, the detritus of the retreating, dying anaerobes.

That change set the terrestrial scene for the emergence of air-breathing multicellular creatures, plants and animals, evolving as the atmosphere became gradually richer, albeit in a fluctuating manner, in oxygen. The chemical turbulence of the Earth's surface settled down in the sense that change became more gradual. Oxygen producers and oxygen consumers, microbial and multicellular, lived alongside each other in a gently fluctuating balance, and the anaerobes went, literally sometimes, underground. Physical stresses such as glaciation, vulcanism, drought and/or inundation continued, sometimes killing off whole species of living things, but reservoirs of bacteria doubtless survived here and there, able either to recolonise devastated areas themselves or, by gene transfer, to enable other types to do so. But sometimes, over geological time, terrestrial and aquatic bacteria could have been subject to extinctions and replacements, perhaps even re-inventions of some lifestyles.

Biologists have recognised for many decades that the activities of living things, specially microbes, have largely determined the chemistry of this planet's surface. And still they do: they sustain the composition of its atmosphere, cycle and recycle the biological elements in its soils and waters, build and erode its rocks; even maintain its temperature. But it works both ways: it is equally true that the bacterial world is conditioned by the rest of the planet, by both its living and its non-living components, and because of their genetic flexibility, bacteria are able to respond to such conditioning with amazing rapidity.

Reflect, if you will, on how drastically mankind has changed this planet's surface during his brief strut on the terrestrial stage. One wonders how much of today's microbial world we have ourselves created.

18

Life's outer reaches

What are the absolute pre-requisites for our kind of life? Obviously the elements that constitute living things must be available in accessible forms: carbon, hydrogen, oxygen, nitrogen, phosphorus, sulphur, potassium, iron and another twenty or so. These elements are abundant in the universe, and are unlikely to be seriously lacking where other requirements of life are satisfied. So what exactly are these requirements? The clearest indications come from the study of microbes, because microbes have colonised the terrestrial habitat to its limits. And they tell us one truth above all others: the features we humans take for granted on Earth's surface today are far from obligatory for life.

Let me recapitulate briefly. Life manages very well without oxygen, evolving into flourishing communities of anaerobes. Acidity, at least to the strength of weak sulphuric acid, presents no problem, as sulphur bacteria and their co-habitants illustrate, nor does a considerable degree of alkalinity bother alkalophiles –

indeed, certain alkaline lakes in Africa hold the terrestrial record for biological activity. Water purity is a trivial matter: saturated salt brines support abundant bacterial life. And pressure is quite as irrelevant, with bacteria growing happily in a near vacuum or at the huge hydrostatic pressures of deep ocean trenches. Temperature, too, presents little problem: boiling hot springs support bacterial life, and bacteria have been found growing at 112° Celsius in superheated geothermal water under hydrostatic pressure; conversely, other types of bacteria thrive at well below zero, provided the water is salty enough not to freeze. And even if they do get frozen, many bacteria revive when their habitat thaws. Even organic food is not a pre-requisite: though organic matter was necessary when terrestrial life first flickered into existence (I told that story in Chapter 14), today plants, and several kinds of bacteria, use sunlight to form their own organic matter from carbon dioxide.

Does that mean that light is essential, then? Sunlight does indeed seem to be a more serious matter. Plants cannot manage without it, and neither animals nor the great majority of microbes can manage without plants. Photosynthesis, by plants and bacteria, provides the primary supply of organic matter on which the rest of the living world subsists. With photosynthesis at the base of our food network, we are all here by courtesy of solar energy. It is true that a specialised group of bacteria exists called the chemotrophs, species of which, instead of using sunlight, can obtain energy from mineral substances such as iron, hydrogen gas, sulphur or ammonia, and use it to make organic matter from carbon dioxide. At first sight chemotrophs might seem to be able to form the basis of a food chain which is independent of sunlight, but in fact they need oxygen gas to attack their minerals with, and this would not be available were it not for plant and cyanobacterial photosynthesis. Chemotrophs depend on sunlight at one remove, so to speak.

Are there any species of microbe which could manage on a sunlight-free Earth? Yes, there are a few. They include certain anaerobic methane-producing bacteria, which need only gaseous hydrogen and carbon dioxide to generate organic matter, and some species of sulphate-respiring anaerobes which can make organic matter from hydrogen and carbon dioxide, given a supply of sulphate. On Earth, today, they are not actually independent of sunlight: the hydrogen these bacteria use comes from organic matter which has been made by other, sunlight-dependent, creatures. But free hydrogen is abundant in the universe, and so are sulphate and carbon dioxide. They *could* manage without sunlight, and that fact means that light energy, though jolly useful to life, is not absolutely essential.

So what are we left with that *is* essential? The answer is water. Though evolution has provided living creatures with ways of living on dry land, they have to keep their interiors wet. Certainly some organisms have developed ways of surviving extreme drought and desiccation, often for long periods, but they do so by becoming dormant, as seeds or spores. They revive only when wetted. The fact of the matter is that terrestrial life processes can take place only in water. It is not just that a liquid is essential for moving molecules around and causing them to interact; the chemistry of water, with its ability to split into its hydrogen and oxygen components, which themselves interact with biological molecules, makes it an integral part of life. Scientists have tried to imagine how life-like processes might carry on in other liquids, such as petroleum, or acids such as sulphuric or formic, or fluids which might occur on colder planets, such as liquid ammonia, liquid nitrogen or liquid methane. But even the most elementary features of terrestrial life's chemistry, let alone the build-up of complex molecules such as proteins and DNA, simply could not happen in any of these fluids.

Water, then, is the fundamental background against which all the processes of life take place. It is an absolute pre-requisite of our kind of life. (I refer to 'our kind of', or 'terrestrial-type', life because I wish to disregard, but not wholly dismiss, the possibility of life-like entities based on parameters wholly independent of our planetary chemistry. Such as sensate stellar plasmas communicating by radio? Yes, I like science fiction, but not just here.) On Earth, the physics of water determines life's theoretical temperature limits: from about $-30°$ Celsius, the coldest unfrozen Antarctic brine, to above $350°$, the temperature of certain deep sea hydrothermal vents. I said theoretical deliberately; in practice it seems that the upper limit is, as I mentioned earlier, much lower at 112°. I discussed why in Chapter 2; it probably has to do with the rate at which biological chemicals concerned with energy transfer are decomposed by heat. Had there been call for it, a more heat-stable terrestrial biochemistry would probably have evolved here.

Elsewhere in the universe, things could be different, for there is no reason to think that the terrestrial limits need be universal: ice melts if it is compressed, and water boils at elevated temperatures; on a high gravity planet, the temperature range of liquid water could be three or four times as wide as here, and the scope for terrestrial-type life would be accordingly greater. (Lifetimes would be short, and life fast-moving, at high temperatures; long and sluggish lives would be the feature of cold worlds. Would those lives feel like ours, with time gently accelerating as one ages? I rather think so.)

Other worlds

What else, apart from availability of liquid water, distinguishes an inhabited planet, such as ours, from one that has no life?

The answer is metastability. This is a word that chemists and physicists use: an object or substance is metastable when it appears stable, but when its stability is maintained only by consuming or conserving energy. A clear example of a physically metastable object is a top which is spinning, because sooner or later the pulse of energy which initiated the spin will run out and the top will fall over. Less obvious is the case in which energy is retained: a top-heavy object which, suitably poised, stays upright until a minute disturbance initiates its collapse – the top, if exceptionally finely balanced, might behave like that when spinning ceased. Another example of physical metastability occurs when pure water in clean glass is cooled steadily without disturbance: it remains liquid for several degrees below its freezing point. It becomes 'supercooled'. Disturb it by shaking, or drop in a solid particle, and it forms ice with a snap. Again a miniscule input of energy is needed to de-stabilise it, but as it freezes it warms up a little: it gives out thermal energy. Actually, glass itself is metastable because it, too, is a supercooled liquid, but it takes tremendous physical stresses to make it crystallise.

Metastability can also be chemical. Explosives are spectacular examples: give them a pulse of heat or a physical impact, as the case may be, and they undergo catastrophic chemical change. And a mixture of cooking gas and air is equally metastable. In a far more gentle way, living things are metastable. They build themselves up, expending energy derived from food or, in the case of plants, from sunlight. But the molecules that compose their functional parts – the cells of their soft tissues rather than structural organs such as shells or wood – are metastable. They could not be truly stable, because living depends on their interacting with each other. So the functional parts of living things have to be kept in continuous repair. Their components are constantly being used up and disposed of, while fresh molecules of the same kind are generated to replace them, all of which requires more energy. Ultimately some critical parts of the replacement processes fail, and then the organism will die.

The living and non-living worlds are so intimately linked that the metastability of living things is reflected in our habitat. The clearest example of such a reflection is provided by the composition of our planet's atmosphere. Today it is rich in oxygen, but 3 billion or so years ago there was none; as I have mentioned in earlier chapters, we have oxygen because plants generate it continuously from water during photosynthesis. Free oxygen is metastable on Earth, a tell-tale indicator of life. Even a billion years ago, when there were neither animals nor plants, and all life was microbial, the atmosphere had a modest oxygen content which would have revealed life's presence. Move back another billion years, when there was no oxygen, and the tell-tale signs of life in the atmosphere would have been hydrogen sulphide and hydrogen: gases which would disappear without microbes to regenerate them. If, through some global catastrophe, life on Earth were now to cease abruptly, free oxygen would vanish fairly rapidly, entering into chemical combination with various terrestrial minerals.

Today's atmosphere has other components which are meta-stable, such as methane: it reacts with oxygen but is constantly generated by microbes. Apart from determining the atmosphere, living things have altered the planet's geology in ways which are blatantly evident: chalk, peat and coal come at once to mind; readers who have dipped into Chapter 8 will know that several ores and mineral deposits are also of biological origin. The biological elements, carbon, nitrogen, sulphur and so on, are cycled through air, water and soil – itself a partly biological product – by living things; forests influence weather and cloud formations. The physics and chemistry of the Earth's present surface together offer a supreme example of the way in which living things, collectively, have not only changed their environment to suit themselves, but sustain that environment in a metastable but steady state, so that it and they change only slowly.

Metastability on a planetary scale is a presumptive indicator of life. Analyses of moon rocks reveal no sign of comparable metastability, nor did anyone seriously expect them to. A couple of decades ago, most of the scientific community was less certain about Mars – except for James Lovelock, a distinguished British environmental scientist, who pointed out that remote sensors, which had sent back reliable information about its atmosphere, had found nothing metastable about it. Therefore the planet was not likely to be inhabited. When the Viking lander of 1976 reached the Martian surface and sent its first signals back, much excitement was generated because, though the atmosphere was just as expected, one part of the probe seemed to have detected life-like processes in a wetted sample of Martian surface dust. But it was a misleading result (of a kind that researchers are all too familiar with): an unexpected chemical reaction between a watery solution provided by the lander and samples of Martian soil had given a false impression. I shall write a little more about life on Mars later in this chapter.

Space explorers in future centuries, human or robotic, will use remote sensing to seek evidence for both water and chemical metastability on or around potentially inhabited celestial objects, long before they or their landers arrive. They will be aware that an inanimate world can mislead. Apart from unexpected chemical reactions that can occur when an alien material is introduced, as happened on Mars, geological activity can simulate metastability. The aftermath of volcanic activity can leave metastable materials in the form of fluid and gaseous emissions which change physically as they cool, or react further chemically; wind erosion can produce bizarre geographical features suggestive of deliberate activity; and fluid movements can sieve, sort and re-shape objects in positively Earth-like ways.

Our space explorers will also be aware of another point. Our dependence on the sun as our primary source of energy has predisposed terrestrial life to occupy our planet's surface, and conditioned most of us to think of planetary surfaces as the only extra-terrestrial habitats. But had our kind of life depended, for example, on the Earth's geothermal heat as its primary energy source, the surface is hardly likely to have been colonised at all. In the last two decades, space probes have revealed a bizarre selection of worlds within the solar system, especially among the satellites of the outer planets. These include Jupiter's Io, a hot, constantly erupting world of sulphur, and Europa, a smooth snow-ball encasing liquid water; Saturn's Titan, with cold seas of petrol-like liquids and non-biological organic matter on a bedrock of ice; and Neptune's Triton, with lakes of liquid nitrogen in a landscape of solid methane. Imaginative space scientists have noticed that Europa, being warm enough to have liquid water, could have developed aquatic, anaerobic inhabitants beneath its crust; they have also pointed out that there is a zone in the vast atmosphere of Jupiter where water is liquid and the sorts of organic compounds from which life emerged on Earth are constantly being formed: buoyant, floating organisms analogous to microbes could exist there. In both instances, such biological systems would be powered by planetary heat more than by the attenuated light of their far distant sun.

After the disappointment of the once promising Mars, floating microbes around Jupiter and anaerobic fish within Europa are but imaginative extravagances – at present. More seriously, is there any likelihood of life in the universe outside the solar system?

Certainly there is. The probability is that we are far from alone.

Water (mostly as ice) and the biological elements are widespread and abundant, and locations in which the elements can come together as organic matter are numerous – radio-spectroscopy has identified many simple organic compounds in interstellar space, and they are definitely present on Jupiter and Triton. There are likely to be millions upon millions of planets in the universe as a whole, and 5 to 10% of them will have liquid water. Even within our own galaxy, the Milky Way, there is probably a vast number of planets, since it is an unimaginably huge conglomeration of between 100 billion and 1000 billion (10^{11} to 10^{12}) stars, and many of these are likely to have planets circulating around them. In 1992 David Hughes of the University of Sheffield estimated that the Milky Way has about 40,000 million (4 $\times 10^{10}$ planetary systems, each of which might include one planet that is temperate and wet enough for terrestrial-type life. Hughes's figure is a huge increase over earlier estimates of some 100,000 (10^5) inhabitable planets, but either number is enormous.

Life may not have taken hold on every wet and temperate planet, and scientists have no real idea of what the proportion might be. But on statistical grounds alone it is likely to be high rather than low, especially because a wet-and-temperate planet need not be very Earth-like to support one or another kind of microbial life (see my opening paragraphs). Airlessness, high or low temperature, saltiness, acidity, alkalinity and high pressure are all tolerable provided liquid water is present. Of course, in such environments evolution is likely to have generated some beings which, by terrestrial standards, are truly exotic. But one thing is certain: where life has moved in, it will have altered the planetary chemistry to suit itself, sustaining local or planet-wide metastability, just as on Earth.

Even if Hughes's number is correct, and even if most terrestrial-type planets are inhabited, there is still no likelihood of such a planet being within 20 light-years of Earth, which means that light or radio signals from here would take two decades to reach the nearest. Over such distances, unless cosmologists' ideas about space and time are spectacularly awry, there is little chance of humanity's interacting with the life forms that have probably emerged on those planets – even if, as is likely, evolution on some has produced beings who are asking themselves similar questions.

Yet if beings at our level of technological development have evolved elsewhere, there is a chance that we could become aware of them. Early in the twentieth century, radio, and later television, were invented, and in consequence the Earth started emanating radio signals, spreading like ripples into interstellar space. They get weaker as they spread, but if a sensitive alien space probe, far out in the galaxy, were to pick up what was left of them 20, 40 or even 1000 years hence, its proprietors would recognise that they corresponded to no ordinary cosmic process, and know that something living was, or had been, active in the neighbourhood of our sun. And the reverse is also true: any alien community which uses radio for communication is unintentionally signalling its existence. That is why NASA has a programme for scanning the skies for unexpected radio signals. NASA has also deliberately beamed radio signals towards the denser part of the galaxy in case they, too, are looking out. After all, most of us would like to know whether there are other beings out there, and no doubt they would like to know of us, too.

Personally, I rather enjoy the thought that the universe probably has other inhabitants, yet I find, to my surprise, that even among scientists there are some who resist the idea of extra-terrestrial intelligences. Their outlook stems from the nineteenth-century naturalist and thinker, Alfred Wallace, who was co-discoverer of natural selection with Charles Darwin. Wallace argued that terrestrial life, and mankind, must be unique in the universe: life was balanced so finely and intimately with the Earth's climate, physics and geology, and even with its position within the solar sytem and galaxy, that a repetition elsewhere of so astonishingly close and delicate a relationship was beyond the bounds of probability. It is a thread of reasoning that persists even today, though rarely among biologists, who have generally perceived its flaw. It disregards the now overwhelming evidence that living things and the planet have evolved together, generating that profound intimacy gradually but spontaneously. We, and now I speak for microbes and plants as well as for my fellow people and animals, have altered the planet to suit ourselves, and if we had not done so, we should not be here. We might be somewhere else (and equally impressed by how well it suited us); more likely, we should not *be* at all; there would be others instead.

However, among non-scientists, the wish to believe that terrestrial life is unique can become almost passionate. With some, this desire stems from a sense that religious dogma is threatened; and I suppose that, since all the world's religions were invented by people, it is understandable that they place mankind at the centre of the universe, and that firm believers wish us to remain there. But in addition, I have been astonished to read of those who feel that humanity would be in some way diminished if other living things were discovered to be sharing the universe with us. And of yet others who truly fear invasion by aggressive aliens in space ships (quite as idiotic as those who yearn for the coming of high-domed savants, who will enforce universal love and peace upon wayward mankind). Well, they are all safe; with interstellar travel out of the question and conversational exchanges restricted to intervals of many decades, perhaps centuries, they need fear no piratical space monsters coming to attack us; no little green men stealing in and out (leaving only corn circles); and no benevolent savants, to belittle us and our gods.

Our world

Mankind has more immediate problems, however. Like a colony of microbes which has suddenly been provided with a glut of nutrient, we are multiplying exponentially: doubling in numbers roughly every 50 years. Our numbers recently passed 5.5 billion, almost spot-on the course predicted by demographers over 20 years ago. This means that by the year 2015 there will be about 8 billion people on Earth. We know this for sure, because the children are already among us who will mate and produce the extra billions, and they will do so before their parents and grand-parents die. Thereafter the numbers become more speculative, but they will certainly go on going up.

Over historical times, our expanding population changed the

appearance of this planet's surface spectacularly and, despite some unfortunate effects such as the deforestation of North Africa, the changes have been greatly to humanity's benefit. But since our population growth became explosive in the twentieth century – 'population explosion' is a very apt description in the time-scale of our history – we have become well embarked on changing things for the worse. Global pollution, global warming, over-fishing, declining ozone layer and such catastrophes have rightly become contemporary buzz-words, though far too many of us choose to forget that the population explosion is the root of so many of these environmental troubles.

Like human beings, microbes often alter their environments in ways that initially favour themselves, but that later become distinctly unfavourable. Unlikely as it may seem (for behavioural analogies between microbes and Man must be few), it is instructive to reflect on what causes a microbial population to cease multiplying.

First, there is the matter of resource depletion. Microbes cease multiplying when a nutrient runs out – usually the principal energy-providing food, but sometimes a minor nutrient such as a nitrogen source or a vitamin. Humans have vastly more complex resource requirements and, despite local shortages, our global habitat is nowhere near exhaustion. As far as food is concerned, our population growth has been sustained by intensive agriculture. Food shortages, even famines and starvation, occur in some parts of the world, but they do not arise from any inadequacy in current agricultural technology – which, according to the late Sir Kenneth Blaxter, could support some 9.5 billion people. They happen because we lack, as a global community, both the political and the economic will to cope. And as far as industrial resources are concerned, we now have sophisticated ways of winning some of them – such as deep mining for coal and offshore drilling for oil – and we use substitutes for resources that have become truly scarce (e.g. using plastics to replace scarce metals).

Our responses to shortages, whether of food or of raw materials, amount to increasing the quantities of energy we have to devote to winning those resources. Technologically it will be perfectly possible to go on doing just that for some time to come, especially if, probably in the late 2000s, hydrogen fusion energy becomes a reality. In principle, we could increase our energy consumption *per caput* sufficiently to feed and supply almost double the world's present population, perhaps more, at present nutritional standards. We have resource problems today, of that there is no doubt; but none is really insurmountable. Not yet, anyway.

However, there is also the matter of the accumulation of damaging wastes. Microbes are especially cogent examples here, because most higher organisms avoid fouling their habitats. But the growth of microbial populations is often limited by the accumulation of toxic end products, such as when the multiplication of yeasts in a fermentation is arrested by the alcohol they have produced, or when the growth of sulphur bacteria is slowed to zero by the sulphuric acid they make. These things happen because microbial habitats are generally closed: the toxic residues cannot escape easily. By this criterion, mankind has joined the microbes – and is already in difficulties. We have become so numerous that our habitat is effectively the whole planet, and is thus closed. Current anxieties over refrigerants (the 'CFCs'), greenhouse gases and global warming, atmospheric oxides of sulphur and nitrogen, detergent and pesticide residues in food and water, radioactive waste disposal and so on, arise because the waste products of human activities are beginning to cause us damage on a global scale – though so far the damage has not been sufficient to limit our population growth significantly.

Thirdly, there is the matter of disease. Epidemics are symptoms of over-population and actually limit the size of populations of higher organisms: a prime example is the effect of myxomatosis on rabbit populations. But any analogy to the biology of microbes *per se* is bland. Bacteria are susceptible to pathogens – to special viruses and to predatory bacteria (called *Bdellovibrio*) – and attacks by pathogens are commonest among dense microbial populations. So it is with people. But microbes are relevant in a way independent of analogies: it is microbes that cause human epidemics, and they are greatly favoured by overcrowding. However, in the last half of the present millennium, medical advances have contained our epidemics and ailments with great ingenuity – the last time mankind experienced such problems on a global scale was as the Black Deaths of the Middle Ages. Today's troublesome scourges, such as malaria, drug-resistant tuberculosis, schistosomiasis and AIDS, are terrible in human terms, but they do not limit our population growth. At their worst they will have barely perceptible effects on global population figures.

Finally, there is the matter of aberrant behaviour, a problem which has no microbiological analogue but which is perhaps the least tractable consequence of humanity's microbe-like multiplication.

Overcrowding predisposes gregarious mammals such as ourselves to aggressive, competitive and anti-social behaviour. If you doubt that assertion, reflect on the behaviour of normally polite and considerate people in rush hours, traffic jams, mass meetings and so on. Such behaviour is precipitated by individuals in the crowd who, giving way to anger or frustration, initiate what amounts to social breakdown among their fellows. Those individuals are examples of the deviants from the social norm who, in all sorts of ways, become the foci of local, regional and occasionally global behaviour patterns, for good or for evil as the case may be. It is self-evident that, because there are more people around than ever before, there are more such deviants than ever before. And those that catalyse social breakdown have, because of more frequent overcrowding, more opportunities to do so than ever before.

We have seen these trends operate impressively during the latter half of this century. Within small communities, there is the familiar progression from vandalism and hooliganism to street crime and violence. Between communities, the situation can become catastrophic: racism, nationalism, or aggressive religious fundamentalism lead to terrorist atrocities and wanton massacre, sometimes erupting into war. The overt causes of such breakdowns differ from case to case. Social analysts might cite poverty, boredom or escapism as causes of localised incidents; dogma, greed or varieties of tribalism beneath more widespread episodes. But they all have deeper biological roots in the response of gregarious mammals to overcrowding. In an ironic way, they are diseases of the twentieth century, because they spread like infections;

diseases which have replaced the microbial epidemics which medical advances have enabled mankind to avoid. Even in Western societies, where population growth approaches zero, there are centres of high population density, the big cities, which are sliding into social breakdown; and outside the West, countries with high population growth rates retain some kind of stability only by means of oppressive regimes, autocratic, oligarchic or theocratic as the case may be – or, if political government is weak, by means of mafias.

It is a dismal spectacle, and biology offers little comfort: it tells us that, if humanity goes on multiplying, things can only get worse. Our descendants will live embattled lives, impoverished by dogma, corruption, crime and oppression.

One is compelled to the conclusion that, even if mankind could find, and use, remedies or palliatives to the environmental and social problems which we are generating, we must still stop multiplying. This is a truth that many otherwise environmentally conscious people still prefer to forget, but there is no getting away from population control. We have only one Earth to live on – at present.

New worlds

Scientists are romantics at heart, even if most keep their fantasies to themselves. The thought that it might be possible to explore the skies beyond the confines of our planet has attracted scientists, as well as thinkers and writers, at least since the era in which the tale of Icarus was first conceived. Exploration has always subsumed colonisation – sometimes regrettably – and as the scope for exploring the Earth's surface dwindled in the course of the nineteenth century, space exploration became an increasingly attractive theme. Throughout the whole of the twentieth century, starting in 1903 with a Russian teacher and writer, Konstantin Tsiolkovsky, there have been a few scientists prepared to write down thoughts about ways in which mankind might colonise extra-terrestrial space. Some favoured artificial constructs: Tsiolkovsky* conceived space stations, and the British crystallographer J. D. Bernal, writing in 1929, imagined huge closed hollow spheres, able to house twenty to thirty thousand people, with animals and plants. Others preferred to consider how to establish settlements on the Moon, Mars and Venus, our nearest planetary neighbours. Such ideas are no longer the stuff of science fiction; the Russians' Mir space laboratory is a reality, in regular use; and feasibility studies carried out in both the USA and the old USSR have provided detailed plans for enclosed settlements on both the Moon and Mars (Venus is agreed to be too hot – though ways of cooling it down have been considered), as well as for constructed space colonies on Bernal's lines.

At the present pace of technological advance, one or more of these projects will probably come to fruition during the earlier centuries of the next millennium. I find it difficult to imagine that any constructed colony in space, or even a few enclosed Lunar or Martian settlements, could do much to alleviate population pressure, simply because of the cost, in energy terms, of transporting from Earth all the installations needed.

But modifying Mars, to make it a self-sustaining environment for unenclosed colonisation, ought not to be beyond the bounds of human ingenuity. It has water, though not much by terrestrial standards, and it has sufficient gravity to retain a breathable atmosphere. But its present atmosphere, mainly carbon dioxide with a little nitrogen and argon, is so tenuous as to be incompatible with human beings, even with oxygen masks (it has less than a hundredth of the density of Earth's atmosphere). Moreover, its temperature ranges from zero Celsius to $-100°$: well below that of Earth's Antarctic. Nevertheless, terrestrial bacteria of various kinds survive well in simulated Martian environments in the laboratory provided they are warmed to above zero. Mars was not always so uninviting. The Mariner fly-past space probes and two Viking landers have provided strong evidence that, around 3½ billion years ago (when terrestrial life was in its infancy), Mars was quite Earth-like. It was sufficiently wet for rain, rivers,

---

* Tsiolkovsky's contributions are not widely known; perhaps it is appropriate that his name is perpetuated as a crater on the *dark* side of the Moon.

flooding and sedimentation to erode and modify its surface, leaving distinctive features that the Mariner probes photographed; it had quite a dense atmosphere, though still based mainly on carbon dioxide and with barely any free oxygen; it was warmer by several tens of degrees; and it had active volcanoes, which would have liberated geothermal heat, lava and gaseous emissions. The more obvious pre-requisites for terrestrial-type life, albeit anaerobic microbial life, existed. But the Viking landers, as I told, provided no evidence that life had taken hold: perhaps the warm, wet period was too short. On the other hand, negative evidence is always tricky. Perhaps life did become established; perhaps Mars is now in the terminal stage of habitation, with the environmental traces of its biology no longer obvious at its surface, yet with anaerobic life forms persisting in residual wet environments beneath its surface.

The water that once splashed around Mars was largely lost into space, together with much of its atmosphere – I must refer you to books on planetary science for the reasons – and today it is a cold, harsh planet, with so exiguous an atmosphere that the little remaining water is in the form of ice or water vapour. Liquid water no longer exists on its surface: with the changing seasons, ice at the poles evaporates direct to water vapour and returns as frost. (This fact gives rise to a paradox: it can be argued that Mars is as wet as it can be, because its atmosphere is actually saturated with water vapour – given its low pressure and sub-zero temperature ranges.) Mars also has quite a lot of oxygen, combined loosely in its surface sands (as compounds called peroxides), as well as more sulphur than was expected.

Can Mars be restored to relative warmth and habitability? Possibly, but I cannot prescribe a step-by-step plan. The question has been discussed in depth among space scientists, and in August of 1991 the science periodical *Nature* published a fascinating survey of the possibilities. One thing we need to know for sure is whether any kind of life persists beneath the surface of Mars and, if so, what to do to conserve it. And sites with liquid water (doubtless as a strong brine) need to be discovered or established. But assuming such a project proves feasible, an essential early step in the 'rehabilitation' of Mars will be the introduction of bacteria of the kind which tolerate Antarctic conditions on Earth, including types capable of generating oxygen through photosynthesis, with the long-term objective of setting up the sort of planet-wide metastability that the Earth's global biology relies upon (including, of course, a spot of our much maligned global warming). Lower plants and simple animals will come later; humans (without space suits) much, much later. Patience will be needed: over a few centuries the planet will have to be managed so as to follow, with due attention to its own geochemistry, transformations which took the Earth over 3 billion years.

But, provided that humanity can contain the stresses of population pressure, and that our world does not degenerate into several hundreds of warring, neo-mediaeval principalities, advancing technology will enable us to know, or learn, what those transformations need to be.

It will be a supreme example of life's ability to extend its outer reaches – as so often, by altering its world to suit itself.

Index